HEALTH, ILLNESS, AND USE OF CARE: THE IMPACT OF SOCIAL FACTORS

RESEARCH IN THE SOCIOLOGY OF HEALTH CARE

Series Editor: Jennie Jacobs Kronenfeld

Previous Volumes in the series:

Volume 11:	Agents of Health and Illness
Volume 12:	Patients, Consumers, Providers and Care Givers
Volume 13:	Health Care Delivery System Changes
Volume 14:	The Evolving Health Care
Volume 15:	Changing Organizational Forms of Delivery Health Care: The Impact of Managed Care and Other Changes on Patients and Providers
Volume 16:	Quality, Planning of Services
Volume 17:	Healthcare Providers, Inst and Patients: Changing Patterns

RESEARCH IN THE SOCIOLOGY OF HEALTH CARE
VOLUME 18

HEALTH, ILLNESS, AND USE OF CARE: THE IMPACT OF SOCIAL FACTORS

EDITED BY

JENNIE JACOBS KRONENFELD

Arizona State University, Arizona, USA

Emerald

United Kingdom – North America – Japan
India – Malaysia – China

Emerald Group Publishing Limited
Howard House, Wagon Lane, Bingley BD16 1WA, UK

Reprints and permission service
Contact: booksandseries@emeraldinsight.com

British Library Cataloguing in Publication Data
A catalogue record for this book is available from the British Library

ISBN: 978-0-76-230740-1
ISSN: 0275-4959 (Series)

CONTENTS

PART II – SOCIOECONOMIC STATUS AND RACE IN HEALTH AND HEALTH CARE

PART III – HEALTH CARE SYSTEM ISSUES: HEALTH POLICY, HEALTH LIFESTYLES AND THE CONSTRUCTION OF PATIENTS, ILLNESS AND DEATH

LIST OF CONTRIBUTORS

Audrey Alforque	California State University Hayward, Department of Sociology, Hayward, CA 94542, USA.
Donna B. Barnes	California State University Hayward, Women's Studies Program, Hayward, CA 94542, USA.
Donna Brogan	Department of Biostatistics, Rollins School of Public Health, Emory University, Atlanta, GA 30329, USA.
Khalil Carter	Duke University School of Medicine, Box 28700, Durham, NC 27708, USA.
Neale R. Chumbler	Institute on Aging, University of Florida, College of Medicine, Department of Health Policy & Epidemiology, 1329 SE 16th Street, Rm. 5128, P.O. Box 100177, Gainesville, FL 32610–0177, USA.
Yvonne J. Combs	Department of Sociology, University of Florida, P.O. Box 117330, Gainesville, FL 32611–7330, USA.
Carla D. Edwards	Department of Sociology, University of Florida, P.O. Box 117330, Gainesville, FL 32611–7330, USA.
Ashley Foster	Department of Sociology, Western Kentucky University, 1 Big Red Way, Bowling Green, KY 42101–3576, USA.
James W. Grimm	Department of Sociology, Western Kentucky University, 1 Big Red Way, Bowling Green, KY 42101–3576, USA.

Joseph G. Grzywacz	School of Health, Physical Education & Leisure Services, 203 Wellness/Recreation Center, Cedar Falls, Iowa 50614-0241, USA.
Barbara Hanson	Sociology-ATKINSON, York University, 4700 Keele Street, North York, Ontario M3J 1P3, Canada.
Marian Katz	Department of Sociology, UCLA, 2201 Hershey Hall, Los Angeles, CA 90095–1551, USA.
Nancy G. Kutner	Department of Rehabilitation Medicine, Emory University School of Medicine, CRM–1441 Clifton Rd. N.E., Atlanta, GA 30322, USA.
Jennifer Leich	Department of Sociology, UCLA, 2201 Hershey Hall, Los Angeles, CA 90095, USA.
Amber McClave	Department of Sociology, University of Florida, P.O. Box 117330, Gainesville, FL 32611–7330, USA.
Beth Mintz	Department of Sociology, University of Vermont, 31 S. Prospect St., Burlington, VT 05401, USA.
Jonathan B. Perlin	Chief, Office of Quality & Performance (10Q), Department of Veterans Affairs, 810 Vermont Avenue, NW, Washington, DC 20420, USA.
Deborah Plechner	Department of Sociology, University of California, Riverside, Riverside, CA 92521, USA.
Kathleen Johnston Roberts	Postdoctoral Fellow and Assistant Adjunct Professor, UCLA Department of Sociology, 264 Haines Hall, Box 951551, Los Angeles, CA 90095–1551, USA.

Ariela Royer Department of Sociology, Indiana University South Bend, 1700 Mishawaka Avenue, South Bend, Indiana 46634–7111, USA.

Michael Schwartz Department of Sociology, SUNY, Stony Brook, NY 11794, USA.

Jay J. Shen University Professor, Division of Health Administration and Human Services, College of Health Professions, Governors State University, University Park, IL 60466, USA.

Thomas T. T. Wan Department of Health Administration, Medical College of Virginia, Virginia Commonwealth University, Richmond, VA 23298, USA.

Philip Williams Department of Sociology, Western Kentucky University, 1 Big Red Way, Bowling Green, KY 42101–3576, USA.

Yi-Hua Xu Department of Health Administration, Medical College of Virginia, Virginia Commonwealth University, Richmond, VA 23298, USA.

INTRODUCTION

Jennie Jacobs Kronenfeld

This volume explores the impact of social factors on health, illness and the use of care. Chapters include both those that focus on patients within the care process and those that focus on providers of care. The social factors that receive the most attention in this volume are sex and gender, race, and socioeconomic status. Some of the chapters explore broader themes of health policy, health lifestyles and what it means in today's health care system to be a good patient or experience a good death. Chapters deal with a wide range of specific health problems including AIDs, various types of cancer, pneumonia and other chronic illnesses. In the range of illnesses and health problems that are discussed, as well as the range of social factors that are considered, these studies represent the current diversity one finds in medical sociology as a field. The volume also includes a range of articles in terms of methodological approaches. While some chapters present broad overviews of the relevant literature in their area and summarize the policy implications of these issues, others use qualitative research methods including detailed interviews to present their issues. Some chapters are written from a feminist social construction position, while others present detailed quantitative analyses of local or national data sets. These different approaches and methods also represent the current diversity within medical sociology as a field of inquiry.

This volume centers on a number of sociological factors quite important in the developmentof medical sociology historically as well as currently. If one examines major texts for the undergraduate course in medical sociology or sociology of health and illness, most would include some attention to social factors as causes of illness and as factors that impact differential treatment that patients receive within the health care delivery system (Matcha, 2000; Weitz, 1996; Weiss & Lonquist, 1997; Wolinsky, 1988). Sections on health behavior and illness behavior are typically an important part of these texts. In various ways, the articles in this volume focus on health and illness behavior and on linkages between patients within the health care system, providers within that system, and broader health care system and policy concerns.

One of the ways that all of sociology has changed over the last 20 years has been by the increased attention to issues of gender. The growth of the journal *Gender and Society* as a sociologically specific study of this topic is indicative of this, as has been the growth of interdisciplinary journals such as *Signs*. Within medical sociology, this attention to issues of gender has also become important over the last few decades. Within health fields more broadly, one major area of discussion the last decade or so has been the under representation of women as the subjects of medical study, as well as the need to increase the numbers of women who are researchers in health care fields. Issues of women and health concerns have also grown in the health literature over the last 20 years. While the close attention to terminology that has long been important in sociology has not been as important in health fields, especially as regards the distinction between the use of gender versus sex as a term, and has resulted in papers in which the terms may be substituted for each other, there are now journals that specialize in issues related to women's health. Since these journals are less sociological in their approach, they less often explore gender as a factor, and focus more on women and health related issues of special or unique importance to women. This volumehas one of its three sections that will focus on both gender issues in health and issues more specific to the health of women.

If one were to review either introduction to sociology textbooks or medical sociology textbooks, in addition to issues of gender, issues of socioeconomic status and race/ethnicity would be prominently included. In some ways, issues of socioeconomic status have formed a core or sociological concerns since the beginning of the field. Issues of race/ethnicity have been a major focus for many years also, predating the interest in issues of gender which became important themes in sociology over the past thirty years. Within medical sociology, most of the major texts now include sections on epidemiology that then discuss issues of social class and race/ethnicity as linked to health outcomes and also discuss the impact of these same variables on health and illness behavior. Over the last decade, most of the new and new editions of medical sociology texts have included chapters on health care delivery, health care policy or health care or health care finance that also talk about the implications of social class and race on these aspects of health and medicine. While focus on issues of race, ethnicity and class have been a major concern of sociologists since the founding of the field, the growth in these issues in the health sciences area is more recent. One of the more vital areas of research in epidemiology today, however, is a debate over the meaning of race as a variable in health studies and some discussion of the importance of race/ethnicity in explaining differences in health status versus the importance of class in explaining differentials in health. Given the vibrancy of concern over these topics, the second section of this volume will focus on the impact of these variables on health and health care, from a sociological perspective.

Health policy and health care system issues have gained greater importance within medical sociology over the last few decades. While medical sociologists have always been interested inissues of what sociology could contribute to changes within health care delivery and health care professions (as reflected in Straus's (1957) terminology of sociology in medicine versus sociology of medicine) as well as to advancing sociological theory and method through studies of health and health related factors, these issues have gained more recognition in basic texts and are important parts of many medical sociologists' research currently. The growth of interests in such topics is reflected in this volume with a group of chapters in the third section focusing on health care system issues.

In this volume in the series, as with previous volumes, a focus on scholarly work in the sociology of health care provides a different type of article than one might find in policy analysis studies conducted for think tanks, evaluations undertaken for health maintenance organizations (HMOs) or state health departments, or reviews and analyses of problems of health care providers written by researchers from within those fields (such as physicians or nurses). Scholarly work in medical sociology typically arises from years of thought, data collection efforts, and data analysis. The work often follows the tradition within sociology of looking at things with a critical eye, with an unwillingness to accept what is already there as the best or most appropriate. Thus medical sociological work often challenges accepted conventions, beliefs and approaches. These articles continue in that proud tradition of work in sociology of health. The research will tie in to a tradition of research in medical sociology and to theoretical concerns that the discipline has addressed for a number of years. This is true of many of the chapters in this volume. They challenge the current health care system, whether as questions about how care is provided to women, how sex categories are constructed in biomedical research, or how good patients or good deaths are constructed..

The chapters in Volume 18 are divided into three sections. The first section includes four studies on Women's Health and Gender Issues in Health Care. The first two chapters focus more on how women receive care and information about diagnoses and deal with issues of women's health, one for a more general health problem and one for a health problem mostly confined to women. The third and fourth articles focus on gender issues more broadly and deal with social constructions of sex categories and feminist perspectives. The first of these articles focuses on women's reactions to being told their HIV antibody test results. Barnes and her colleagues place the issue of HIV testing for women within the framework of lower rates at this point for women for positive HIV status. Because the issue is less common for women, and because many of the women are dealing with other serious social issues such as poverty, drug rehabilitation and being a single mother, it is important they be told these results in a context that encourages further contact with the health care delivery system. Issues of class, perhaps more than issues of race/ethnicity or gender, end up being an important influence on how women are told their results. This study provides an important illustration of the impact of gender as well as the impact of other social factors considered in more detail in the second section of this book.

The second chapter examines the social correlates related to diagnosis and treatment for breast cancer. After reviewing the existing literature on breast cancer screening, diagnosis and treatment, the authors identify social and demographic factors which help explain the decision making process. As with the first study in this section, this links not only issues of women but issues of income, education and race/ethnicity to the types of care people receive. Poorer, less educated women and women from minority racial and ethnic backgrounds may receive fewer options for treatment than then influence their decision about what to do and even their outcomes from thedisease.

The third chapter by Hanson uses a feminist social construction argument to raise questions about the use of sex categories in biological research. Content analysis of medical writing about cancer since 1900 provides the data that enrich the discussion in this chapter. The author argues that dichotomous sex categories reflect social pressures to categorize and create difference. The scheme is overlaid on continuous physical phenomena. The fourth article also incorporates a feminist perspective, and applied that critical perspective to a consideration of medical sociology and issues in women and medicine. This presents a broad, partially historically based perspective to these issues. It points out that during the second wave of the women's movement, feminists focused on having attention brought to women's health issues. This effort has had some successes, partially as reflected in the growing numbers of women who are physicians. However, medicine and its view of issues in many ways remains highly gendered. The chapter illustrates this through a consideration of the medicalization of women's lives, including in areas of reproductive health. As with much of the best work today in medical sociology, this article concludes the section on women and gender issues by arguing that there is a need for a critical and feminist sociological understanding of medicine.

The second section of this volume has chapters on socioeconomic status and race in health and health care. Three of the four studies include some examination of socioeconomic status and health and health behaviors, sometimes in combination with another factor such as race. The last chapter in this section combines an examination of gender, sexual orientation and race/ ethnicity. Three of the four papers also examine these topics as linked to a specific health problem, such as HIV/AIDs, catastrophic health problems later in life or pneumonia. The fourth relates socioeconomic status tomore general health behaviors. In the first chapter in this section, Shen and his colleagues examine the impact of socioeconomic status on pneumonia. As with many infectious diseases today, social issues linked to this disease have not received as much attention in the literature as have chronic illnesses, even though pneumonia is the sixth leading cause of death in the U.S. and the number one cause of death from infectious diseases. This paper uses data from hospital inpatient discharges and is one of the most quantitatively sophisticated chapters in the volume. It points out both sociological and policy relevant conclusions, as it demonstrates substantial variation across socioeconomic status in the patterns of seeking care and the outcome of treatment for pneumonia. It also demonstrates the importance of public health insurance programs, especially Medicare, as a safety net as well as the need for more care for the uninsured with these types of problems.

The second chapter by Grzywacz focuses on socioeconomic status and health behaviors among a sample of Californians. It disaggregates the association between different indicators of socioeconomic status and then examines the association between these factors and health behaviors by gender. Different hypotheses exploring the linkages between socioeconomic status and health behaviors are examined. Overall, educational attainment had the most robust associations with health behaviors, although different patterns of association between indicators and health behaviors emerged in multivariate analyses. The article discusses implications for sociological thought about socioeconomic status and also for interventions that target lower status individuals with goals of changing health behaviors.

Linkages between race and socioeconomic status are also explored in the paper by Kutner and Brogan. Chronic renal failure, also known as end-stage renal disease, is the health problem under study in this chapter. It also employs more sophisticated quantitative research methods,especially race-specific logistic models. Socioeconomic status is often a significant factor, and race and socioeconomic status are often linked in complex ways, which the chapter tries to explain. It also makes suggestion for future research in this area.

The fourth study in this area uses different methodological approaches and also tries to explain how patients view providers of care and how this interacts with social factors, a different focus than the other three studies. In-depth interviews were conducted with 28 HIV positive patients. Over half of patients expressed preferences regarding physician gender, almost half expressed a preference about physician sexual orientation, but only a fifth expressed preferences regarding physician race/ethnicity.

The last section of the book includes five chapters on health care system issues. Articles present arguments about health policy, health lifestyles and managed care. The first two both deal with issues of construction of appropriate roles for patients, one in the outpatient setting and the other in the way to die. They both look at how health care providers view good patients and good deaths. The first chapter by Katz discusses constructing the good patient in conventional and unconventional medical settings. It uses data segments from doctor-patient symptom reporting interactions to determine what is viewed as a "good patient". Differences between interactions styles and conventional versus unconventional settings of care are discussed. The paper by Leich explores the relationship between hospice home care and dehospitalization trends. It explores the dilemma of having hospice home nurses who coordinate and help plan care for dying patients also be responsible for fiscal issues for their employers. Does the task of facilitating the wishes of their clients as to an appropriate death conflict at times with the task of managing care in a cost efficient manner for their employers?. While many of the nurses do value death at homeas an option and explain that to family caregivers, it is also often in the financial interest of their employers as well.

The third chapter in this section by Mintz uses a macro-perspective to examine health policy issues. This explores the role played by corporate capital in the development of the U.S. health care system. Capital is defined more broadly than in some other research, to also include public expenditures and subsidies. The development of the U.S. health care system has been driven by the interaction between the publica and private sectors of care. Government spending, therefore, may have as important an effect on corporate development as does legislative action.

The last two chapters in this section both deal with aspects of health lifestyles, but in very different ways. Chumbler and his colleagues investigate the influence of mid-life adult statuses such as gender, marital status, parenthood, presence of children in the home and employment and functional health status on selected health related behaviors. Behaviors studied include preventive medicine behaviors, risk-taking behaviors, and medication usage. Data to explore these issues come from a local telephone survey in Kentucky. From this study, the authors derive several testable, theoretical propositions that can be used in future research to test the relationships between these factors. Royer explores issues of special medical care needs of those with chronic illnesses. She argues that social as well as medical management of services needs to be considered. Uncertainty is viewed as a key characteristic of chronic illness and one that causes many problems for both the chronically ill person and people delivering care to that person. Although uncertainty is in many ways a non-medical characteristic, in this study, it was mentioned as more important than medical difficulties, financial burdens or frustrations with health care providers. While uncertainty is a different type of social factor than such measures of status as socioeconomic status or race, orconsiderations of issues of gender, this study helps broaden the understanding of social factors to include other types of variables as well and thus provides a thoughtful paper to conclude this volume.

REFERENCES

Matcha, D. A. (2000). *Medical Sociology.* Needham, Massachusetts: Allyn and Bacon.

Straus, R. (1957). The Nature and Status of Medical Sociology. *American Sociological Review, 22,* 200–204.

Weitz, R. (1996). *The Sociology of Health, Illness and Health Care: A Critical Approach.* Belmont, California: Wadsworth.

Weiss, G., & Lonnquist, L. E. (1997). *The Sociology of Health, Healing and Illness.* Upper Saddle River, New Jersey: Prentice-Hall, Inc.

Wolinsky, F. D. (1988). *The Sociology of Health-Principles, Practitioners and Issues* (2nd ed.). Belmont, California: Wadsworth.

PART I

WOMEN'S HEALTH AND GENDER ISSUES IN HEALTH CARE

"LIKE I JUST GOT A DEATH SENTENCE": CONDITIONS AFFECTING WOMEN'S REACTIONS TO BEING TOLD THEIR HIV ANTIBODY TEST RESULTS AND THE IMPACT ON ACCESS TO CARE

Donna B. Barnes, Audrey Alforque and Khalil Carter

ABSTRACT

HIV and AIDS case surveillance continues to reflect the disproportionate impact of the epidemic on racial/ethnic minority populations, especially women, youth and children. The HIV antibody test is the standard method for identifying people with HIV and the primary prevention model to promote treatment and the reduction of HIV transmission. Given the incidence of HIV in women, it is important to understand the conditions under which women receive and interpret the results of their positive HIV antibody tests, in order to promote access to and continued health care. Our study demonstrates that how individuals are told the results of their HIV antibody test and what pre- and post-counseling is given, if any, can influence individuals' medical care and prevention of transmission of HIV to others. The important issue, it seems, for post-test counseling and accessing HIV health care is to address women's multiple social issues of

Health, Illness, and Use of Care: The Impact of Social Factors, Volume 18, pages 3–33.

ISBN: 0-7623-0740-4

poverty, drug rehabilitation, if needed, shame, and stigmatization of women with HIV, especially HIV positive mothers. Most women lack HIV education; they need referrals and assistance accessing and continuing health care, unless they have stable health care relationships. Analysis suggests, though the sample is too small to conclude, that issues of class, perhaps more than race/ethnicity and gender, though these are analyzed as interconnected, influence how women are told their results and their subsequent reactions and actions. Our study suggests that there may be little relationship between mandated state policies and counseling and testing practices. Policy sets standards. But how policies are implemented can be dependent on multiple conditions including counselors' training, resources of time, HIV knowledge, social support, knowledge of gender, race/ethnicity and social class, and women's social, cultural, emotional and physical health needs.

I. INTRODUCTION

There is a widely publicized reduction of HIV infection rates in the overall population, though the reduction has been less in women, people of color and the poor (Centers for Disease Control, 1998, 1999). HIV and AIDS case surveillance continues to reflect the disproportionate impact of the epidemic on racial/ethnic minority populations, especially women, youth and children (Centers for Disease Control and Prevention, 1998). "Between, 1995 and, 1998, the incidence of AIDS was reduced by 38% in men but only 20% in women – making gains among women only about half those among men" (Gollub, 1999: 1479).

The HIV antibody test (CDC, 1986, 1987) is the standard method for identifying people with HIV and the primary prevention model to promote treatment and the reduction of HIV transmission. "Testing and counseling is the first step to treatment" (Hunter, 1994: 36). Research has examined the effectiveness of HIV testing and counseling in relationship to sexual behavior outcomes (DiScenza, Nies & Jordan, 1996; Grinstead, 1997; Hirano, Slaughter, Sands & Gellert, 1994; Kamb, Fishbein, Douglas, Rhodes & Rogers, 1998; Landis, Earp & Koch, 1992; Sikkema and Bissett, 1997; Weinhardt, Carey, Johnson & Bickham, 1999; Wolitski, MacGowan, Higgins & Jorgensen, 1997), though the majority has primarily focused on gay or bisexual men. The limited research on behavior outcomes of women at risk for HIV includes a much cited study examining women at risk receiving negative test results. "It appears that past behavior is a better predictor of current behavior than receiving counseling and a negative test result" (Ickovics, Morrill, Beren, Walsh & Rodin, 1994:

447). A review article of thirty-five research projects reported on twelve studies examining HIV testing and counseling for women. The studies primarily examined pregnancy among women and heterosexual couples and provided mixed findings (Wolitski et al., 1997). A more recent review article of twenty-six studies indicated that people who learned they were HIV positive did reduce their sexual risk behavior, nonetheless the article did not differentiate between male and female behavior (Weinhardt et al., 1999).

Given the incidence of HIV in women, it is important to understand the conditions under which women receive and interpret the results of their positive HIV antibody tests, in order to promote access to and continued care. With the potential use of protease inhibitors (Markowitz et al., 1995; Collier et al., 1996) the need for recognition, early intervention treatment, and continued contact with providers for people with HIV/AIDS has increased. Niemcryk et al. (1998) indicated that the most consistent predictor for continued HIV service utilization is previous use of that service, more than race, education, and insurance. Gender was not a factor in predicting continued HIV service in this large sample population which included eighteen percent women. Understanding the process of receiving positive HIV antibody test results from women's perspective is important to identify the conditions that influence their accessing social service and medical care programs, allowing early intervention, continued treatment and reduction of HIV transmission.

In this study, we examined, from HIV-positive women's perspectives, how they were told their positive HIV antibody test results, what information they were given, and their reactions to this information. In addition, we examined the state-mandated HIV antibody testing policies, and in what ways state policies and clinical counseling practices are related.

II. THE PROBLEM

A. HIV Testing and Counseling for Women

According to Weissman et al. (1995), women received pre- and post-HIV antibody test counseling at a significantly lower rate than men did. In addition, more men than women were advised to obtain medical treatment following their first positive HIV test. This study suggests that women may not fully understand the information they receive during counseling. Another aspect of differential HIV care based on gender is providers' perception that HIV is not a major risk to women, while in a survey of obstetricians-gynecologists, half (54.5%) reported that unprotected vaginal intercourse with a seropositive partner was associated with high, as opposed to moderate or low, transmission risk (Boekeloo, Rabin, Coughlin, Labbok & Johnson, 1993). Phillips, Morrison, Sonnad & Bleecker (1997: 174) discuss the confusion of providers over current recommendations for HIV testing for pregnant women, and conclude that "providers may be missing opportunities to encourage women to be tested."

The insufficient research on HIV positive women's counseling needs suggests there are limitations to current counseling practices and strategies, and recommendations are made for exploring new approaches (Campbell, 1995; Gollub, 1999; Heise & Elias, 1995; Russell & Smith, 1998). Gollub (1999: 1480) declared that "the actions (frequently but inappropriately reduced to the term "behavior") required to block the spread of HIV cannot be divorced from the context of sex inequality and its ensuring dependencies, power imbalances, and threats of violence – in short, the effects of women's status worldwide as second-class citizens."

B. Sociology of Gender and Health

In the last decade, feminist theorists studying gender have maintained the importance of examining the interconnectedness of gender, race/ethnicity and class (Baca Zinn & Dill, 1994; Collins, 1991, 1994; Green, 1990). It is not enough to add women to the existing theories of gender, which place male dominance at the center of analysis. Collins (1991: 225) suggests that "placing African American women and other excluded groups in the center of analysis opens up possibilities for a both/and conceptual stance, one in which all groups possess varying amounts of penalty and privilege in one historically created system." A concept Collins (1991: 225) calls "matrix of domination," is an interlocking system examining gender, race, ethnicity, class, age, sexual orientation and religion and their oppressive effects on an individual's everyday social world. Studying the matrices, the interlocking systems of aspects of social forces, demands the avoidance of preconceived categories, and the re-examining of taken-for-granted assumptions about an array of social conditions. To attempt this Smith (1987: 10) suggests beginning "with women's experience from the women's standpoint and explore [ing] how it is shaped in the extended relations of larger social and political relations."

Fisher (1986: 132), in her book, In the *Patient's Best Interest*, challenged the assumptions that medical decisions are based on scientific knowledge and that physicians will act in their patients' best interest. Based on analysis of physician and patient interactions, she argued that the relationship is characterized by: (1) physicians' decisions which flow from value-laden social perspectives, as well as scientific information, and (2) "as long as medicine is practiced for profit by primarily white middle-class men, there will be conflicts of interest rooted in economics as well as in race, class, and gender." Fisher argued that a contextual web or interrelational phenomenon exists linking social interaction between the physician and patient and the social structure.

Based on taped interactions of physicians and patients, Waitzkin (1989) suggested that patients present personal problems to their physicians which may often be based in social issues. However, the physician usually tries to find ways in which the patient can individually adjust to her or his personal problem, placing little or no emphasis on social issues. Mishler (1984) similarly argued that physicians dominate the medical encounter with their medical discourse which tends to cut off any conversations about social issues by directing conversation toward technical topics. When patients raise personal problems related to social issues, physicians typically interrupt or otherwise steer the topic back to medical issues.

Danziger (1986) described various types of relationships between physicians and female patients, from a paternal to a more co-participant approach. She argued that the physician has more power in controlling the amount of information given, because the encounter takes place not on neutral territory, but at the physician's office, and patterns of deference are most likely unchallenged by the majority of patients. The author suggested that the gender difference of the male doctors and the female patients heightened the paternal approach to information-giving, as well as increasing a sense of the vulnerability and need for social support by women. West (1984, 1993) found that male doctors tended to interrupt patients more often than female doctors. She suggested that social control by male physicians reflected patterns of social control within wider social contexts. The question still remains as to whether varying configurations of gender, class and race of the physician and women patients would change the type of relationship of medical encounters from paternal to co-participant, and from asymmetrical to egalitarian.

C. Sociology of Gender and HIV/AIDS

Treichler (1988: 232–3) challenged feminists to analyze the HIV/AIDS epidemic exploring the intersections of gender, race and class in relation to the biomedical discourse when meanings suggest nonintersecting categories. Osmond et al. (1993: 115) in a paper entitled, "Multiple Jeopardy Of Race, Class, And Gender For AIDS Risk Among Women," found the challenge a struggle.

> We want to express what Collins (1990, 1991) and Hooks (1981) call the simultaneity of oppression. Our language, however, is not one that combines but, rather, one that

> categorizes and dichotomizes. Thus, with race, class, and gender, we struggle with terms that do not satisfactorily express experiential reality, such as cumulative, multiplicative, interacting, interlocking, intersecting, interpermeating, and so on.

A method utilized by some feminist researchers examining gender, race/ethnicity and class is to be part of the process and not separate themselves from the women they study (Lather & Smithies, 1997). Lather and Smithies (1997: xvi) doing health care research with, not *on* women with HIV/AIDS, illuminated ways to be part of the research.

> This work has made a claim on us to not drown the poem of the other with the sound of our own voices, as the ones who know, the "experts" about how people make sense of their lives and what searching for meaning means.

The study redefines women with HIV and contributes to the deconstruction and reconstruction of women living with HIV/AIDS. However, the sample of twenty-five women with HIV/AIDS was primarily white ($n = 16$), and better educated than the nationally reported cases of AIDS.

Other studies (Hudson & Morris, 1994; Murphy, Mueller & Whitman, 1996; Osmond et al., 1993; Quinn, 1993) examined HIV/AIDS and women of different ethnic groups. Of these studies, Quinn (1993: 309) alone discussed specific ways in which race, gender, and social class contributed to the risk of HIV infection among African American women. "However, it should be noted that race is not itself a risk factor." Quinn's analyses demonstrates that HIV/AIDS is but one of many social, economic and health issues. Depending on available resources to African American women, other issues – such as decent education, employment, child care, health care access, and health care without prejudices – may be more relevant to general well-being than HIV/AIDS. "Whereas the lack of educational and economic opportunities are tragic, the loss of health opportunity is unforgivable" (Quinn, 1993: 312).

A key issue for this study is examining the relationship between state policies and actual HIV antibody counseling and testing practices for women. Clarke and Olesen (1999: 3) argue that despite years of feminist research, the biomedical model of health care is presumed the most common framework for analyzing health care and policies today.

> This model centers on concepts of health status, health behaviors, and technoscience interventions. Gendered, cultured, historicized, classed, raced, and otherwise situated women are routinely silenced or erased as actors in the production of health, in both the provision and receipt of health care per se as well as in health politics and policy.

The social, economic and political conditions that influence policy may or may not be compatible with women's health care needs. According to Hunter (1994: 49), "the intersection of gender, race, and class is nowhere more dramatic than in the law's impact on women with HIV disease." Schneider (1992: 21) emphasizes the primary importance of race, class, and gender and HIV/AIDS health care needs:

> Race, class, and gender are the three social factors most determinant of a person's health status and his or her degree of well-being. In concert, they will affect perceptions of health and illness, kinds and availability of care, modes of delivery, anticipated illnesses, and discourse and interaction patterns of doctor-patient relationships.

There is limited empirical research on women's reaction to being told their positive antibody test results, and how that reaction affects their behavior in seeking and continuing health care. Women need to access effective HIV/AIDS education, social service and medical care programs in social settings with providers that acknowledge that HIV/AIDS is but one of multiple social, economic and health issues for women. Without the understanding of the interconnectedness of race, class, and gender, and the social context within which women and women of color negotiate social behavior and relationships in the context of HIV/AIDS, it is seriously difficult to effectively understand most HIV positive women's health needs (Schneider, 1992: 28).

In this study, we examined, from HIV-positive women's perspectives, how they were told their positive HIV antibody test results, what they were told, and their reactions to this information. In addition, we examined the state-mandated HIV antibody testing policies, and in what ways state policies and clinical counseling practice are related.

III. METHODS

The study was conducted from 1995 to 2000 and is part of a larger project on HIV positive women's reproductive decision-making processes. Using a convenience sample, seventy HIV-positive women were recruited by community and social workers, from social service agencies serving people with HIV/AIDS, in Oakland, California ($n = 24$), Chicago, Illinois ($n = 20$) and Rochester, New York ($n = 26$).[1] To assure integrity in the recruiting process, the first author (hereafter referred to as the P. I.) trained the recruiters to broadly select women in regard to inclusion criteria of age, race/ethnicity, and HIV-infection. The inclusion criteria were female, age 18–50 years, all races, ethnicity, and social class, a self-reported positive HIV antibody test result, and a health stage that allowed the women to participate in a one-hour interview. The exclusion criteria were men, women under 18 years and over 50 years, at terminal stages of AIDS, and no reproductive experiences. At the scheduled interview, the P.I. further selected participants on inclusion criteria of reproductive experiences broadly defined as engaging or not engaging in sexual behaviors that can lead to pregnancy, ideas and desires about conception, strategies related to birth control, experiences of child birth, miscarriage, stillbirth, and/or abortion subsequent to learning HIV antibody positive test results. The recruiters were trained to avoid revealing inclusion criteria to prevent potential participants from trying to fit study criteria for the payment reward. Interpreter assistance was used for women who had limited or no English.

The women were introduced using first names only to the P.I. who completed the face-to-face interviews of approximately one hour in length in a private space, most often in the library or meditation room at the community agencies. Some women from Chicago were interviewed at the P.I.'s private office which was part of a university medical and social service complex. All facilities were familiar locations to the participants. Potential participants were informed that the purpose of the study was to investigate how women with HIV/AIDS make reproductive decisions, and that they would be asked to respond to questions about their HIV/AIDS experience. One potential participant did not meet the inclusion age criteria, and was not interviewed. Participation was voluntary, and a twenty dollar honorarium for women's participation and expertise, and round trip bus tokens were offered. After reviewing the written informed consent, which was also available in Spanish, participants gave verbal agreement to volunteer for the study, and their signature, using first name only on the receipt for cash was written consent to participate in the study. The research protocol was approved by the Institutional Review Board of California State University, Hayward.

Oakland and Rochester were selected as mid-sized cities comparable in HIV/AIDS incidence and are both close to large cities that are epicenters of HIV/AIDS. Chicago is a bigger city and different from the other two cities in setting. Most of the study sample contracted HIV from injection drug use or as sexual partners of men who use injecting drugs, and they represent women with HIV from the three states which are major injection drug-use-states.

Half (56%) of the sample were African American ($n = 39$), followed by white ($n = 15$), Latina ($n = 9$), native American ($n = 3$), Asian/Pacific Islander ($n = 2$), and African (2). Over half (59%) of the sample had some high school education ($n = 20$) or had graduated from high school ($n = 21$). Slightly over two thirds (70%) had an annual income of less than $14,999 ($n = 49$), which is considered below poverty line (U.S. Census Bureau, 2000). The women ranged in age from 18 to 50 years, with a mean age of 36. More than half (64%) reported currently being in relationship with a significant other ($n = 34$) or were married

($n = 11$) with partners who were primarily HIV negative ($n = 30$), and the others were HIV positive ($n = 13$). Other women ($n = 23$) reported being single, divorced, separated or widowed.

The sample represents exposure categories of heterosexual sex ($n = 52$), injecting drug use ($n = 21$), and receipt of blood transfusion ($n = 2$). Some women reported more than one category of exposure. The participants self-reported their disease status by the date of their first HIV positive antibody test, latest T-cell count, and current symptoms of HIV/AIDS. Fifty-five women reported a disease status of HIV positive, and fifteen women reported having AIDS.

For the purposes of this study, the questions asked related to women's reasons and decision to have an HIV antibody test, their experiences of having the test, what pre- and post-counseling they recollected, how they were told about their results, and by whom, their reactions to this information, to whom, if anyone, they told their results, and what action they took after learning their test results; for instance, where did they go, what did they do, that day, and some days and months thereafter, and what feelings did they experience as a result of the positive test. The interviews were audio-recorded, transcribed, and analyzed.

Grounded theory qualitative method of inquiry was used to analyze the interview data (Strauss & Corbin, 1998). Qualitative research is most suitable for process-based studies and allows for participants' self-reflection of actions and motives (Beardsell & Coyle, 1996; Sobo, 1995). Techniques of various levels of coding, memo writing, and diagramming were used in order to inductively build substantive theory from empirical evidence. Potential concepts from initial interviews became hypotheses which were deductively compared and contrasted with additional interview data. The subsequent phases utilized alternatively inductive and deductive reasoning through a reciprocal relationship between data collection, analysis, and concept building employing a constant comparative technique. Emerging concepts and eventually substantive theory were challenged and further developed with the assistance of grounded theory consultants (Strauss & Corbin, 1998), and later in the analysis stage were verified with community outreach workers, social workers, and some participants in the study (Bloor, 1997). In this method, instead of establishing a thesis to be tested in the field, "the researcher begins with an area of study and allows the theory to emerge from the data" (Strauss & Corbin, 1998: 12). The research approach allows what is essential to the phenomenon of women's HIV antibody testing experiences and subsequent actions to emanate as a social reality from the women's perspective.

Table 1. Demographics of Study Sample
Women's Reactions to Being Told Their HIV-Antibody Test Results
Chicago, Illinois, Oakland, California, and Rochester, New York.

Characteristic	Study Sample ($n = 70$)
Age	No.
18–25	4
26–35	27
36–45	34
over 46	5
Mean age = 36	
Ethnicity	No.
African American	39
White (Non-Hispanic)	15
Latina	9
Native American/Alaskan Native	3
Asian/Pacific Islander	2
African	2
Last Grade Completed	No.
Less than high school	20
Graduated high school	21
Some college, trade school	15
Graduated college	9
Some graduate education	5
Annual Household Income	No.
Less than $10,000	40
$10,000-$14,999	9
$15,000-$19,999	2
$20,000-$24,999	3
$25,000 or more	15
Don't know	1
Current Relationship Status	No.
Significant other	34
Married	11
Single, divorced, separated, or widowed	23
Missing data	2
Partner's HIV Status	No.
HIV positive	13
HIV negative	30
Other	2
Not applicable	23
Missing data	2

Table 2. HIV Status
Women's Reactions to Being Told Their HIV-Antibody Test Results
Chicago, Illinois, Oakland, California, and Rochester, New York.

Study sample ($n = 70$)	
Exposure Category for HIV infection*	**No.**
Injecting drug use	21
Heterosexual contact*	52
Sex with injecting drug user	21
Sex with bisexual male	6
Sex with person with hemophilia	1
Sex with HIV-infected person, risk not specified	22
Blood transfusion	2
HIV Status at Time of Interview	**No.**
AIDS	15
HIV positive, asymptomatic	55
T-Cell Count at Time of Interview	**No.**
Less than 50	1
Between 50 and 200	12
Between 200 and 500	29
Greater than 500	20
Don't know	8
Year of First Positive HIV-Antibody Test	**No.**
1986	3
1987	3
1988	3
1989	6
1990	8
1991	9
1992	6
1993	7
1994	6
1995	5
1996	2
1997	4
1998	2
Unsure	6

*some participants gave more than one response.

The unstructured interview data was entered into the computer-assisted program, Non-numerical Unstructured Data Indexing Searching and Theorizing (NUDIST), for data management (Gahan & Hannibal, 1998). Demographic data from the structured interview was double-coded and double-entered into SPSS (Statistical Package for Social Sciences). Study participants' identification was kept separately by the P.I., and codes are used in the following report to protect the anonymity of the women.

IV. FINDINGS

The findings will be presented in four sections. First we will describe the three states' policies and representative women's experiences of being told their positive results. Second, we will compare state policies and women's experiences of antibody testing in three cities on how and what they were told and their reactions to this information. Third, we will report women's action, focusing on their experiences of accessing health care. Fourth, we will present conditions that promote women's access to health care based on women's experiences.

A. Mandated State HIV/AIDS Counseling and Testing Policies and How Women Are Told Results

In this section we will present California, Illinois and New York mandated state HIV counseling and testing policies, with examples of women's experience of HIV testing from these states. First we will describe California's mandated HIV antibody test policy followed by a representative Oakland woman's experience of being told her positive test results ($n = 24$). Then we will describe the same for Illinois, followed by the experience of a woman from Chicago, ($n = 20$), and finally New York's state policy and the experience of a woman from Rochester, ($n = 26$).

California's Health and Safety Code pertaining to HIV tests, passed in 1985 and amended in 1995, mandates that "no person shall test a person's blood for evidence of antibodies to the probable causative agent of AIDS without the written consent of the subject of the test" (Section 120990a). These HIV-antibody tests "shall be made available . . . on a confidential basis" (Section 120895a). Before the test, subjects must be "informed about the validity and accuracy of the antibody test" (Section 120895a). During post-test counseling, medical and emotional referrals must be given, but social support is not stipulated. Patients must be given a "referral list of physicians and surgeons or clinics knowledgeable about AIDS . . . for further information and explanation of the test results for medical evaluation . . . and referral for 24-hour inpatient and mental health services" (Section120895a). In addition, California law states that HIV-antibody test results must be given in person. Although information about preventing transmission is not required, California awards contracts to early intervention projects. Among other objectives, these projects are designed to "provide information and education, including behavior change support, to HIV infected persons to prevent the spread of HIV infection to others" (Section 120900).

A 27-year-old African American woman from Oakland learned of her HIV positive test while in jail, and seven months pregnant.

> I just happened to be in the doctor's office and she just asked me would I like to take an HIV test. I didn't decline because I never thought it could be me. So I took the test, and I'm the one who was very promiscuous and I was a hooker and I had drug problems and stuff. She [female physician] took a deep breath and she just told me I was HIV positive. And I was just like coming out of this cloud; I was just detoxing. I'd been detoxing like for months because I was on drugs so long and everything. I just felt like my body was still detoxing. So I was getting ready to have this baby, hearing my baby's heartbeat, knowing I'm going to have a clean baby and everything, won't be on drugs and then to find out I was HIV positive. It just kind of just shattered me. I can never forget the moment she told me that I was positive and stuff. I felt like I was going to faint and I was crying. It was a woman physician and I had became very fond of her from the time that I was in there. When I started crying she just left out the room. I just kind of went to pieces and I went and talked to the social worker there and I went and called my Mom and told my Mom and my sister (1010).

She said the physician was "not exactly helpful." It is not clear whether this woman received the information and referrals mandated by law from the social worker.

Illinois' Public Health Statutes about HIV tests, passed in 1987, mandate that "written informed consent of the subject of the test" (410 ILCS 305/4) must be obtained before an HIV antibody test is administered. The test is confidential; any "subject of a test who wishes to remain anonymous shall have the right to do so" (410 ILCS 305/6). The test must be explained, "including its purpose, potential uses, limitations and the meaning of its results" (410 ILCS 305/3). Before ordering an HIV-antibody test, the physician must give "the person tested information about the meaning of the test results, the availability of additional or confirmatory testing, if appropriate, and the availability of referrals for further information or counseling" (410 ILCS 305/5). This statute does not explicitly require any post-test counseling.

A 29-year-old African American woman from Chicago went to a hospital for a pregnancy test and told us about the follow-up appointment.

> The next appointment that I had, I spoke to the physician. She came and told me that I was HIV positive, and I was like, I didn't take the tests for HIV. I was just being tested to be pregnant (2005).

Some women in this study did not give consent for their HIV-antibody test and unknowingly were tested for HIV. This particular woman went on to explain how the physician reacted when questioned about the lack of consent for the test.

> She [the physician] didn't say anything. She didn't say anything. So I sat in the office, and then she came back, she asked me was I okay. I was like, "No". I had to sit in the hospital for like 2 or 3 hours; I'm in shock. I was pregnant and being all HIV positive also (2005).

Post-test counseling was not apparently offered or available at this hospital. When asked if anyone came into the room to check on her, or give her HIV information, she replied:

> Nah. Just that last person said, "Are you okay?" The one that gave me the appointment recommended I go to another hospital for the HIV. I didn't go back to the hospital until I had my baby. After that I went out and used drugs. I remember when I went on a really wild binge (2005).

This woman did not receive prenatal care or HIV information, medical or social services. She went to the emergency room of the same hospital, gave birth to a son who tested HIV positive at birth, but seroconverted to HIV negative by 16 months of age, at the time of the interview.

New York's Public Health Laws referring to HIV tests, passed in 1988, mandate that "no person shall order the performance of an HIV related test without first receiving the written, informed consent of the subject of the test" (Article 27-F, Section 2781, 1). The test is confidential and "confidentiality protections" must be explained to the subject of the test. The test must be explained "including its purpose, the meaning of its results, and the benefits of early diagnosis and medical intervention" (Article 27-F, Section 2781, 2a). During post-test counseling, the subject must be given

> counseling or referrals for counseling; (a) for coping with the emotional consequences of learning the result; (b) regarding the discrimination problems that disclosure of the result could cause; (c) for behavior change to prevent transmission or contraction of HIV infection; (d) to inform such person of available medical treatments; and (e) regarding the test subject's need to notify his or her contacts (Article 27-F, Section 2781, 5).

New York's law is comprehensive; it requires medical, emotional, and social service referrals as well as information about preventing transmission.

From Rochester, a 35-year-old Puerto Rican woman, through a translator, said she was not given information during post-test counseling. She was tested while in jail and depended on fellow inmates for HIV/AIDS prevention information. From Spanish to English, the social worker translated the P.I.'s question of whether the woman was given any counseling and her answer.

> The next question you asked regarding how did they give her any counseling, pre- or post-counseling, she said not really, that it was just a physician that gave her the results and just explained to her that she has to take AZT, that her T-cell count was very low. But regarding transmission prevention, she did not receive any counseling. Really what she learned, she said was from one of the inmates, that they talked, amongst the inmates (3001).

She was given medical care, but no information about preventing HIV transmission.

These representative examples demonstrate similarities in women's HIV counseling and testing experiences such as the lack of HIV information including prevention of HIV transmission, and lack of counselors' emotional support and referrals for social services. However, examination of the state policies demonstrates some differences.

B. Comparison of State Policies and Women's Experiences

California, Illinois, and New York each have policies regarding HIV antibody counseling and testing. These policies include requirements for the information and referrals that must be given to individuals before and after HIV antibody testing. The policies can have many components including information about consent, anonymity, the HIV antibody test, the virus, the disease, and transmission prevention. Referrals can encompass medical and social services. The counseling and testing policies of these three states differ in the way they address these components.

The pre-counseling required in the three states is similar in that each state stresses the need for consent, confidentiality, and an explanation of the test; however, post-test counseling varies. New York's law mandates that clients must be informed of "behavior change to prevent transmission or contraction of HIV infection (Article 27-F, Section 2781, 5), which neither Illinois nor California law requires. California does have early intervention projects that give information about preventing further transmission of HIV. Although these projects are awarded contracts, they are not mandatory. California law does state that results of the HIV test must be given in person, but this is not stipulated by the other two states. In fact, Illinois' law does not expressly state a requirement for post-test counseling. Additionally, California's policy differs from those in Illinois and New York in that it provides reimbursement to counties that "submit a plan . . . that details where testing and pretest and posttest information and referral will be provided and the qualification of the staff who will be performing the services" (Section 120895)

There are three kinds of policy contexts, respectively in California, Illinois and New York. Within these differing contexts we compared women's perspectives on receiving their positive HIV antibody test result, hereafter referred to as positive test results, examining patterns and conceptualizing themes. Three interconnected themes emerged: (1) women's multiple social conditions, (2) their perceptions of HIV positive test results as a death sentence, and (3) the influences of such perceptions, social conditions, and HIV counseling and testing experiences on women's accessing medical care and social support services.

For women with an annual income of less than $14,999, ($n=49$) which represented slightly more than two-thirds (70%) of the women in our study, what and how they were told their HIV test results compounded the multiple social situations they already faced with limited knowledge and resources for change. These included discrimination based on gender, race/ethnicity and class, poverty, homelessness, drug use, and/or a drug-using environment, and prostitution. A 33-year-old African American Chicago woman was tested when she went to her doctor for a two-week check up after her third baby. She describes her former life as a drug addict and prostitute.

> I jumped in and out of cars . . . I walked the streets three, four, five o'clock in the morning by myself, jumped into strange cars with strange men, had sex in abandoned buildings, got high in abandoned buildings, in cars, just did everything. I got raped a couple of times, at gunpoint, at knifepoint, and came out of it OK. It's just kinda like a miracle, you know (2008).

A 50-year-old African American woman from Oakland told of being tested for HIV in the state penitentiary. She was released before receiving her test results. She said:

> So when I got out I went and got on a methadone program to make sure I stayed out. And I participated in the groups there. They had lots of people talking on AIDS, nurses coming down there talking to groups. And I realized that some of the symptoms they were saying, that I had some of these symptoms. And so I asked where can I get tested at and they sent me to a clinic. And that's where I went and got tested and I went back about six weeks later. They had a list of a lot of people. There was about 50 people there to get their test results and you just went in one at a time, one at a time. And when it was my turn, they just said, "Next please" and I went in there and they asked me for my piece of paper with my number on it. She went out and she came back about 5 minutes later and she said, "You're positive," and "Next please." That was it. She didn't tell me nothing else. She didn't tell me what to do. She didn't tell me where to go. She was like I just got a death sentence (1004).

This was in 1986, one year after California mandated policy to give a "referral list for AIDS medial care . . . further information and explanation of the test results for medical evaluation . . . and referral for 24-hour in-patient and mental health services" (Section 120895a).

Women's experiences of receiving tests results were hearing or interpreting, or being made to feel, that the HIV positive test results were a "death sentence." Women in this study recalled being told that their test results were positive, that is, they had the HIV virus, not AIDS. However, the majority of them subjectively interpreted that information as "You Have AIDS." They equated the results to a death sentence. This was universally true regardless of the city they lived in, or the mandated policies of the states.

> A white woman from Rochester said, "I just thought immediately I was headed for a death sentence" (3026).
>
> An African American woman from Oakland with three children said, "I felt like I was going to die right there, you know. I became very hysterical and laid on the floor and I was just waiting to die then" (1013).
>
> A 36-year-old white woman from Rochester said, "And the first think I thought of is 'I'm gonna die" (3018).
>
> An African American woman from Chicago said, "they told me that I was HIV positive. I told her [my mother] that I had AIDS, because I thought it was the same. I assumed it was AIDS. I just assumed it was a death, a death . . ." (2005).
>
> A white woman from Oakland said, "I thought I was given a death sentence. The only woman I knew with AIDS was dying" (1028).
>
> An African American woman from Chicago said, "I was devastated. I was dead after I got up. I had a baby that was one, and my little baby in my arms and he sit there and I'm broke and he sit there and he told me I was a junkie and I was just like, the only thing on my mind was that I was gon' die" (2008).

Some women, believed they were going to die because they lacked the knowledge to know the difference between being HIV positive and having AIDS. Others knew they were HIV positive, but their own experience of knowing people who had AIDS, whom they had observed wasting away and dying of AIDS, overshadowed what they were told. For many women the HIV positive test results added to the burden of multiple social conditions, leaving some with few resources of medical care, social services, and emotional support from family, and friends.

A smaller group of women heard the positive HIV test results but tried to live as if they were not HIV positive. "I had just erased it" (2016). "I was very shut down, you know. All I wanted was everybody to leave me alone" (1013).

> I guess I had just basically blocked it out. It was like I didn't have it. It was like I was living life like I was the same person. I didn't believe that I had it. I was just blocked it out. And I went on with my life (1020).

Even fewer women questioned the testing procedure and/or the integrity of the counselor. A 36-year-old African American woman from Chicago said, "I went to the clinic because I was having problems with my stomach. I had an enlarged liver. I had a drinking problem" (2006). The health care provider asked her if she would take an HIV test, and she consented.

> Then the test results came back along with the test results from the other things that were going wrong with me. They came back positive. I didn't believe it. I told them they were lying, they had my [HIV] tests results mixed up with somebody else's. I mean, I didn't listen, and I walked out of the clinic (2006).

Like some women, her reaction to the positive results were to relapse into drug use.

> I go to rehab[ilitation] and I got all my stuff together, and I left. I was in treatment, and I left. I went back home and I got high. I got drunk first, real drunk. Then as time went on, I started getting high again (2006).

All three states have policies that include requirements for information and referrals that must be given to individuals who have the HIV antibody test. However, there is the potential for counselors' not following state-mandated policies, resulting in women not being given information and referrals. There is also the potential for mis-communication between two people for a variety of reasons including women hearing and interpreting the results incorrectly, being in shock, not believing their results, leaving before the counseling session is complete, or being unable to process an additional condition to survival in their everyday life.

As we have seen, most women's reactions to their positive HIV results were largely perceived as a death sentence, possibly influenced by multiple social conditions and experiences, with HIV becoming just one more oppressing condition, and potentially the more fatal condition. Also some women attempted to ignore the information, and fewer women did not believe the test results.

We now transition from how women were told, and their reaction to the information, to women's actions. How does being told positive HIV results and women's reaction to this information affect their accessing HIV medical and social services?

C. Women's Action and Accessing HIV Health Care

In this section we will focus on women's actions after learning their positive HIV results. We are particularly interested in the conditions under which women access HIV medical care and social services. The states' policies, implemented optimally, could assist women to access health care by providing HIV information and referrals to social service and medical care programs. Yet in practice many women were not given the benefit of those policies, or for some other reason were not able to make use of health care resources.

How women are told their HIV antibody test results, including within a health care setting, influences and can postpone early intervention. In the following example a 43-year-old African American woman from Chicago went to the hospital for a miscarriage, and to her recollection, was tested for HIV without her permission. A physician told her the test results.

> He told me like it, just matter of factly, you know. He just told me. And he didn't ask or wait for me to have any questions or anything. He told me and left. And that was the worst day of my life for a long time you know. I was devastated. I didn't know what to do. I was afraid to tell anybody I had it 'cause I felt that they would ostracize me and for a long time after that I was in deep denial for years, you know. Maybe if I had came, if I had had somebody to talk to me, uh, I might have started getting cured sooner you know, but I didn't (2003).

The lack of compassion and knowledge can exacerbate an already traumatizing situation. It can also discourage women from accessing needed medical care and social services.

A 32-year-old African American woman from Oakland, was told she had AIDS, when she was actually HIV positive. She was a nurse's assistant and a mother of five children, two of whom are HIV positive. She described her situation.

> I guess he was a supervisor, and he told me that I had AIDS. And, of course I was shocked . . . it was kind a like a real mystery to me, 'cause I thought that was for only like . . . gay men had got the virus and stuff. I didn't know anything about it. But he didn't say, that I had HIV, just that I had AIDS. I went to a public Hospital, and that's when I found out about HIV, it's the virus that causes it. And I didn't have AIDS (1020).

This case was the only instance in our study of a woman being given an inaccurate diagnosis. Her lack of HIV/AIDS knowledge, however, was common to many women in the study. She did not receive referral for medical care, however she pursued treatment in another medical facility.

The interaction between information given and how it is presented, influenced how and where women accessed health care. A 46-year-old African American from Rochester said,

> She [physician] told me, and I was quite upset. And she told me, you know, that I wouldn't be able to come back to the hospital for treatment, that she would send me somewhere where it was more professional and what was wrong with me. So I felt kind of bad about that because I had been going to that doctor for a long time. After she told me that, I wouldn't say she hated me, but after that was told to me it was a change so that made me

feel real bad. And I just wondered why, like hurry up and get rid of me as a patient? (3008).

The woman's reaction to the lack of compassion influenced her to avoid accessing information, medical treatment or social care from that agency. After two months of secrecy and isolation she reached out to a family member, a cousin who was HIV positive and he helped her get care.

Drug use and the shame attached to it contributed to some women's perceptions and actions. Some women in this study after learning their positive test results continued using drugs, others relapsed to drug use, and others continued potential risk behaviors for HIV transmission, of sharing contaminated needles for injecting drugs, and unprotected sexual intercourse.

A 39-year-old white woman from Chicago learned she was pregnant and HIV positive at the same time in 1991. She stopped using drugs and prostituting when her son was born. She recalls her lifestyle, experience of being tested, and her action:

> I was an IV drug user and eventually I ended up working on the streets to make money for my drugs. One day I woke up and it seemed that my boobs had grew over night. So I went to what we called the AIDS clinic and they tested me for pregnancy and HIV at the same time. Well, I tested positive for both. She [counselor] was very understanding and very concerned. Like I said, for a long time they tried to get me in there. And a couple of times I went in there but if I had to wait longer than five minutes, I was back out the door. I know much more now, than I did then. I'm not really sure if AZT at the time that I was pregnant was known. She could have explained it to me, and at the time my frame of my mind was . . . I might not have heard it even if she had (2019).

Knowledge of protocols such as Zidovudine 076 (Connor et al., 1994) might not have been available, or if available, might not have been heard by some women.

> My immediate response was that I would terminate the pregnancy. It was explained to me. Because at that time I did not know that distinction between HIV and AIDS. I didn't. I just assumed that all babies born to HIV positive mothers would be born HIV positive. At the time, I thought that even if my child was born negative I would probably die. You know I thought of a lot of things, but then I just took the drugs so I wouldn't have to think about anything (2019).

Some women sought and received knowledge of HIV prevention from multiple sources such as friends, inmates, family, counselors and medical and social support providers. A woman from Chicago said, "I began to read, and it was like drinking in knowledge" (2013). And some women utilized this knowledge to prevent HIV transmission to others. However other women did not have this knowledge, or if they did, did not use behaviors that prevented HIV transmission

Women with HIV have spoken of shame and the fear of being shamed and stigmatized by providers and have avoided medical and social services. Women with HIV who are using drugs, or in recovery from drug use are particularly vulnerable to stigmatization and shame. Guilt for being pregnant also influences isolation and can prevent women from seeking prenatal care. A 31-year-old African American women with HIV, who is also a mother and outreach worker in Oakland, said,

> I didn't seek prenatal care, for a lot of reasons. I was afraid the system would be against me. I was afraid they'd say, "Why did you get pregnant, knowing you're HIV positive?" I was happy about being pregnant and I was afraid they'd try to make me have an abortion because of the HIV. I had been raped, and I was still using crack. I was afraid of being judged (1029).

A common response of keeping HIV a secret leads to isolation. Some women found social support groups that offered them HIV knowledge, emotional support and brought them out of isolation. A former injecting drug user and prostitute from Chicago introduced earlier recalls after three years of secrecy about her HIV status, her experiences in support group.

> I started seeing people there that I knew. It wasn't such a shameful thing no more. And I saw women that were beautiful, intelligent women. You know and they were suffering from this too. Most people think ah, it's that type, that old greasy, nappy head women that's got this virus but it's not just her, you know (2003).

There are many reasons, and many conditions why women may not access care, in spite of help in some cases from social workers and providers. One woman described the efforts of a nurse to guide her into treatment but she still held to her conviction that she was dying.

> Right away my assumption was I'm going to get AIDS and die and she [nurse] explained to me that that's not necessarily true. But as much as she explained to me I thought well I don't know anyone who has not been HIV positive who did not get AIDS and die or is dying (2019).

The overall picture of treatment of the majority of HIV positive women in our study illustrates the importance of a number of social conditions that discourages women from accessing care. Starting from the almost universal assumption that HIV leads to death, and the lack of adequate information on HIV/AIDS, some women additionally are not given referrals for care. Others suffer from drug addition and are unable to meet responsibilities for their own treatment or they feel ashamed of their lifestyle, test results and/or pregnancies and avoid exposing themselves to others, including providers. Others have had discriminatory health care experiences in the past, or have known others in their communities and families who have died of AIDS. For all these reasons women avoid seeking treatment after HIV positive testing. The question is under what conditions do women who are told they are HIV positive access health care?

D. Conditions That Promote Access To Health Care

Although many women reported a scarcity of compassion, information, and referrals when being told their test results, other women described counselors and providers who were compassionate. The previous examples are differentiated by income from the following examples of representative women's experiences. The previous examples had an income below $14,999 ($n = 49$). The following examples of women have an income of $25,000 or more ($n = 15$) representing 21% of the study, and have more education than the former group and/or have professional or occupational status.

A 40-year-old Asian mother of a 13-year-old son is a chemical engineer from Rochester and was in relationship at the time she decided to have an HIV antibody test.

> I was considering splitting up with my significant other at the time, and which I did eventually. But I kept thinking, if I dated someone I would really feel badly if I didn't know because I did have quite a few risk factors. And I really was very shocked that it came back positive because here, for years, I mean, between 1983 and 1991 is when I found out, I was running three miles a day. I was the healthiest I had ever been in a decade. Oh, I thought for sure that meant I was going to die, I mean, even though I know better (3010).

She possessed HIV knowledge of risk behaviors and prevention of HIV transmission prior to being tested. She also had access to regular health care and knew how to direct her medical care. After she learned her test results, she took her son to the physician for an HIV antibody test, challenging the physician's advice.

> My son was healthy. Actually, that was the first thing I did was to get him tested, although the doctor said, "I don't really think we have to do that," and I said, "Well, you never know" (3010).

A 37-year-old woman from Rochester had left her 11-year-old son with her mother in Africa, when she came to America to work for the African Ambassador to her country. She was studying for her immigration test, and needed the HIV antibody test in order to get her green card to continue to work in the United States. She had been married for 6 months to an American. He accompanied her for the HIV antibody test. She tells what happened when she learned her positive HIV results.

> The doctor gave me the number to call for some help. I was like, oh, no, I just want to die now. He said, "No, you can't die. You can live. This is the number you can call and they will help you" (3005).

Before she left, she understood from the physician that she should use condoms to prevent HIV transmission to her husband. She lied to her husband about the test results. She was afraid to tell her husband for fear of retaliatory violence. But as soon as he left for work, the same day she learned her results, she called the number she had been given, and was seen within 3 days at an HIV community-based medical and social service agency. Her husband accompanied her to the first appointment and the social worker asked the woman in private, "Do you want me to tell him that you are HIV positive?"

> Yes, of course, that's why I'm here because it was hard for me to just bring it out in the house. I didn't know what he was going to expect of me or just throw me out or whatever. I'm scared (3005).

The social worker assured her that "if he does that, we will keep you. Don't worry anymore, because you are in good hands, now." And she called the husband in and told him.

The Director of an HIV/AIDS agency in Oakland is another example of an HIV positive women being told her positive HIV test results and effectively accessing health care. She is a 38-year-old white woman with two Bachelors degrees, by her own admission two more than one needs to run a non-profit agency. She is in a 15-year relationship, married and has twin 4-year-old daughters who are HIV negative, as is her husband. She was enrolled to begin law school when she learned her positive HIV antibody test results. She had taken the HIV test to support and accompany her girl friend, who was very frightened to have the test. The Director's first thought when being told she was HIV positive was, "I'll never be able to have children." My second thought was, "Have I infected my husband?" and my third thought was, "How can I tell my mother" (1028). (1028) was an immigration paralegal, working for two lawyers, and an activist for immigration justice at the time. Like most women she thought she had been given a death sentence. The only woman she knew with AIDS was dying. But she knew how to get information. She became an AIDS activist almost immediately. She asked questions of the counselor who gave her the test results, and began networking with women, health care workers, participated in clinical trials, accessed the best available health care for herself, and started a newsletter for other HIV positive women.

Another Oakland woman, a 29-year-old African from Nigeria, was studying nursing at a prestigious University and like her husband of eight years was sponsored by the YMCA. She had a very high fever and went to her private physician for care, who gave her an HIV antibody test. She recalls, "I remember the doctor couldn't figure out why I had a fever so high" (1006). She describes her experience of being told:

> He [the physician] kind of beat around the bush. He didn't want to just straight out tell me. He tried to comfort me before he told me, and asked me how am I feeling? Am I okay? And then he told me that I was HIV positive. He said you don't have AIDS, but you have HIV. I just sat there. I went home and just looked at my husband. I just sat there and started crying. And the doctor called me again two days later and he knew I was married, cause he's known me, that was the same doctor I always used. And so he asked me had I told him [the husband]. He told me to come in and get a blood test to see if he [the husband] caught the virus, and I said I hadn't told him. I didn't know how to tell him. I said would you tell him for me. I don't know what to do. Can you tell him and he did. He was a very polite doctor, very nice, and very understanding. That's why I asked him to tell my husband (1006).

This woman expressed that she felt the physician gave the results to her personally, instead of his nurse who administered the test, because he knew her. When I asked if her physician had questioned her about potential risk behaviors, such as injecting drug use, she said, "No, he knows me. He knows I didn't use drugs. That's why. Because he knows me" (1006).

From the data, there was one exemplary case of post-counseling told to us by a middle class, 41-year-old African American woman from Chicago. She was educated at a private liberal arts college and is an editor and writer. She had the HIV test because her former boyfriend in another city had died of AIDS. Mutual friends called to inform her. The writer remembers the moment when she was told her test results.

> Not a moment I'll ever forget. It was a counselor, a Latina. She was pretty good, I mean in retrospect, she really was pretty good, very empathetic, and very compassionate. I was so positive that I was not positive, that, I went alone, to get the results. I thought, I'm in great health. And I had all of these, um, notions, mistaken assumptions so, I had not at all read about the test. I just assumed, um, negative. So she told me very gently. She said remember I said to you few weeks ago, that you should try to take some time during these few weeks to talk to people you really trust, so that you can be prepared for whatever the outcome is. And I wasn't even really listening to her, but I heard her say those things. And she told me that the results were positive. And I just went into shock. I don't really remember what happened for several seconds. I was just totally in shock. But she stayed with me for several moments, and she said, she asked me who was with me. She gave me a list of places to call and she stayed there with me until I made some contacts and follow up with someone. I thought that was typical, but when I talked to other people later, I found that's not at all typical. People just let you go. You're just out there. She was very careful about referrals for health care, and social support and she just kept going over those details with me and, and see, the first three calls I made, no one answered. So she could have just said well she's on her own. But, no, she said, you can't leave unless you have the contact made. So I did make contact with an agency here, with someone who became my counselor for three years. And, later became really good friends. Before I left, she made sure that I made a contact with a personal friend that I had told I was getting tested 'cause she was concerned that I might go home and have a gap of time before I saw the counselor that I made contact with. She was very, she was like an angel to me, cause when I left there, I knew where I was going. I went to my friend's office, and she went there with me. And then

> about a half hour after I got home, the HIV agency counselor called and set up an appointment for the next day. So I had a really critical support family that kept the floor from falling out (2011).

What these representative examples of women's experiences have in common is middle class social status, defined by income, education, social position and/or occupation.

Like the other women in the sample, the middle class women also felt like the positive tests results indicated a death sentence, but unlike the other women, they had resources of skills, knowledge, stable health care to access HIV health care. Some of these women, like the former examples, were former injecting drug users, had been battered, and/or had had sexual intercourse with bisexual men or men who used injecting drugs. Often the women did not know this information until they had positive test results and had inquired of former boyfriends, their families and friends.

These women also shared race/ethnicity with the other women in the sample, but had not had discriminatory experiences in health care settings, and their health care was generally in private practices, health maintenance organizations and not in public hospitals. A 38-year-old middle class African American woman from Rochester shared her comments about social class differences.

> The women that I know, that I've sort of networked with, that have this virus. My lifestyle is so different from a lot of these women. Then, there's a group of women, that our lifestyle is very similar. But, from bein' part of the support groups it's that thing of neediness that I get from certain uhm, women. Of, uhm, needing that love, needing – it's, so it's, it goes really, really deep. There's a lot of soul wounds. There's a lot of hurts (3021).

This group also had different motives for seeking their HIV antibody test, for instance helping a friend, following procedures for employment, or anticipating new relationships.

It appears conditions which influence women's accessing HIV medical care and social services are: (1) middle class status defined by one or more of the following: income, education, and/or occupational status, (2) knowledge and skills to acquire knowledge, (3) stable medical care, usually without discriminatory experiences due to race, or perceived drug use, and (4) referrals for HIV health care.

V. DISCUSSION

This presentation focused on how women were told their positive test results, their reactions to this information, and their actions, focusing on accessing health care, with possible relationship to state HIV counseling and testing policies. Our study built on Treichler's (1988) work attempting to analyze social issues related to HIV by examining the intersections of gender, race and class within a biomedical discourse.

Our study demonstrates that how individuals are told the results of their HIV antibody test and what pre- and post-counseling is given, if any, can influence individuals' medical care and prevention of transmission of HIV to others. The important issue, it seems, for post-test counseling and accessing HIV health care is to address women's multiple social issues of poverty, drug rehabilitation, if needed, shame, and stigmatization of women with HIV, especially HIV positive mothers. Most women lack HIV education; they need referrals and assistance accessing and continuing health care, unless they have stable health care relationships. Analysis suggests, though the sample is too small to conclude, that issues of class, perhaps more than race/ethnicity and gender, though these are analyzed as interconnected, influence how women are told their results and their subsequent reactions and actions. In addition, stigmatization associated with how women contract HIV, i.e. injecting drug use, sex with injecting drug users, and prostitution, may also have a discriminatory affect on the counselor-client interaction and influence women's reactions and actions. Building on Collins' (1991) theory, our study indicates empirically that these interlocking conditions have their oppressive effects on individuals' everyday social worlds.

A study of this size has limitations and is not generalizable to other than similar samples of HIV positive women and can not be used to predict behavior outcomes. Rather, this qualitative study offers a close examination of the process, patterns and conceptualization of women's perceptions of HIV counseling and testing. Future studies examining counselors and the influences of their characteristics, ideologies, gender, race and class on the client-counselor interactions could offer further directions for HIV policy and counseling. Further research utilizing a larger study sample examining the influences of social class, and exposure categories of HIV and counseling practices could prove enlightening.

Our study builds on the research of Otten, Zaidi, Wroten, Witte and Peterman (1993) that suggested perhaps there are problems with the HIV counseling procedures, and that the lack of risk-behavior reduction after counseling could be a function of inadequate counseling. Our research agrees with Wolitski et al. (1997: 65) that HIV counseling and testing is not "equally successful in influencing behavior across a wide range of settings and populations." Various methods of counseling are still being debated including increasing the amount of counselor-client time (Hirano et al., 1994), utilizing a format of short but personal and interactive counseling (Kamb et al., 1998), and counseling to reduce high risk sexual behavior which may lack any significant change in clients' HIV/AIDS knowledge (DiScenza et al., 1996). We are not offering suggestions for counseling format, though there is evidence which can offer direction. Rather we stress the importance of recognizing the influence on health of social and political contexts as well as gender, race and class, building on others' research (Clarke & Olesen, 1999; Danziger, 1986; Fisher, 1986; Waitzkin, 1989; West, 1984, 1993). While taking into account the mandated polices our research demonstrates that the setting, resources, counselors' characteristics, stigmatization for women's HIV exposure categories and pregnancies, women's multiple social needs, and women's motive for testing are all conditions of the dynamics of HIV counseling.

Our analysis indicates that each state's policies differ in important ways. The primary difference in the three states' HIV antibody testing policies was Illinois' limited, if any, mandatory post-counseling. Yet the exemplary post-counseling case was found in Chicago, and we acknowledge programs that provide more than the mandated state policies for HIV positive women in Chicago (Cook County Hospital Women & Children HIV Program; Chicago Women's Aids Project, Midwest AIDS Training & Education Center). Our analysis further suggests more similarities in women's experiences of HIV antibody counseling and testing, regardless of the city and state in which they received their HIV test results.

Our study suggests that there may be little relationship between state policies and counseling and testing practices. However, state-mandated HIV policies can set minimum standards and enforce them, and the state can address women's HIV health care needs by allocating resources that have meaning both for services available to women and sources of support, both economically and socially, for counselors. Policy sets standards. But how policies are implemented can be dependent on multiple conditions including counselors' training, resources of time, HIV knowledge, social support, knowledge of gender, race/ethnicity and social class, and women's social, cultural, emotional and physical health needs. We strongly support post-counseling that is marked by compassion, empathy, and understanding of women's needs, current and accurate HIV information, and referrals for social support and medical care, ideally with contacts made during the post-counseling session. With the potential use of protease inhibitors the need for both early intervention and continued contact with providers for people with HIV has increased. Post-counseling which initiates contact with health care providers and promotes long-term and continued relationships between provider and clients will contribute to the reduction of HIV transmission.

ACKNOWLEDGMENTS

The authors wish to thank the consultants, Sheigla Murphy, Diane Beeson, and Susan Taylor-Brown for their contributions to data analysis and generous feedback on earlier drafts of the manuscript; the research assistants Jessica Tello, Ethel Gamboa and Tina Duncan for their assistance in the project, and Sandy Boucher for editing. The P.I. wishes to thank her daughter, Monica Bill Barnes for her comments on drafts of this and previous work and her continued belief in women's legacies. Thank you to the staff at the community social service agencies for their support and recruiting efforts: Dorothy Chapman, Lyn Blackburn, Zulma Ortiz, Yolanda Newman, Carol Smith, Joan Palmer, Mary McCauley, Sheri Goult, Joanne Despotes, and Mary Lewis. The P. I. also wishes to thank the women who shared their experiences and their reflections with me. Your stories have now been told to a wider audience. This research is funded by National Institute of Health Grant 3 SO6 GM/A148135–04S1; University of Illinois at Chicago Fellowship, Center for Research on Women and Gender, funded by John D. and Catherine T. MacArthur Foundation; Research Grant, California State University, Hayward.

NOTE

1. In Oakland the women were recruited from a community based social service agency for women and men with HIV. In Chicago the majority of women were recruited from a recovery program for HIV positive women. In Rochester the women were recruited from a community based medical and social service agency for women and men with HIV. In Chicago as compared to Oakland and Rochester there was a greater incidence of injecting drug use after the women were told they were HIV positive. This might be explained by a greater number of Chicago women who were injecting drug users before their HIV tests, and therefore the Chicago sample may be biased towards recovering injection drug users.

REFERENCES

Baca Zinn, M., & Thornton Dill, B. (1994). Difference and Domination. In: M. Baca Zinn & B. Thornton Dill (Eds), *Women of Color in U.S. Society.* Philadelphia: Temple University Press.

Beardsell, S., & Coyle, A. (1996). A Review of Research on the Nature and Quality of HIV Testing Services: A Proposal for Process-Based Studies. *Social Science and Medicine, 42,* 733–743.

Bloor, M. (1997). Techniques of Validation in Qualitative Research: A Critical Commentary. In: G. Miller & R. Dingwall (Eds), *Context and Methods in Qualitative Research* (pp. 37–50). London: Sage.

Boekeloo, B. O., Rabin, D. L., Coughlin, S. S., Labbok, M. H., & Johnson, J. C. (1993). Knowledge, Attitudes, and Practices of Obstetricians-Gynecologists Regarding the

Prevention of Human Immunodeficiency Virus Infection. *Obstetrics and Gynecology, 81*, 131–136.

California (1995). *California Codes, Health and Safety Code, Sections 120990, 120990a and 120895a*. Retrieved October 1999 (http://www.leginfo.ca.gov/cgi-bin/waisgate?WASdocID=6917519572+5+0+0&WASaction=retrieve).

Campbell, C. A. (1995). Male Gender Roles and Sexuality: Implications for Women's AIDS Risk and Prevention. *Social Science and Medicine, 41*, 197–210.

Centers for Disease Control (1986). Additional Recommendations to Reduce Sexual and Drug Abuse-Related Transmission of Human T-Lymphotoropic Virus Type III/Lymphadenopathy-Associated Virus. *Mortality and Morbidity Weekly Report*, (March 14), 152–155.

Centers for Disease Control (1987). Recommended Guidelines for HIV Antibody Counseling and Testing in the Prevention of HIV Infection and AIDS, (April 30) Appendix IV, Mimeo.

Centers for Disease Control (1998). *HIV/AIDS Surveillance Report, 10*(2), 37.

Centers for Disease Control (1999). *HIV/AIDS Surveillance Report, 11*(1), 18–19.

Centers for Disease Control and Prevention (1998). *National HIV Prevalence Surveys, 1997 Summary*. Atlanta, GA: Centers for Disease Control and Prevention, 1–25.

Collier, A. C., Coombs, R. W., Schoenfeld, D. A., Bassett, R. L., Timpone, J., Baruch, A., Jones, M., Facey, K., Whitacre, C., McAuliffe, V. J., Friedman, H. M., Merigan, T. C., Reichman, R. C., Hooper, C., & Corey, L. (1996). Treatment of Human Immunodeficiency Virus Infection with Saquinavir, Zidovudine, and Zalcitabine. *The New England Journal of Medicine, 334*, 1011–1017.

Clarke, A. E., & Olesen, V. L. (1999). *Revisioning Women, Health, and Healing*. New York: Routledge.

Collins, P. H. (1991). *Black Feminist Thought: Knowledge, Consciousness, and the Politics of Empowerment*. New York: Routledge.

Collins, P. H. (1994). Shifting the Center, Race, Class, and Feminist Theorizing About Motherhood. In: M. Baca Zinn & B. T. Dill (Eds), *Women of Color in U.S. Society*. Philadelphia: Temple University Press.

Connor, E. M., Sperling, R. S., Gelber, R., Kiselev, P., Scott, G., O'Sullivan, M. J., VanDyke, R., Bey, M., Shearer, W., Jacobson, R. L., Jimenez, E., O'Neill, E., Bazin, B., Delfraissy, J-F., Culnane, M., Coombs, R., Elkins, M., Moye, J., Stratton, P. M., & Balsley, J. (1994). Reduction of Maternal-Infant Transmission of Human Immunodeficiency Virus Type 1 with Zidovudine Treatment. *The New England Journal of Medicine, 331*, 1173–1180.

Danziger, S. K (1986). The Uses of Expertise in Doctor-Patient Encounters During Pregnancy. In: P. Conrad & R. Kern (Eds), *The Sociology of Health and Illness: Critical Perspectives* (pp. 310–21). New York: St. Martin's Press.

DiScenza, S., Nies, M., & Jordan, C. (1996). Effectiveness of Counseling in the Health Promotion of HIV-Positive Clients in the Community. *Public Health Nursing, 13*, 209–216.

Fisher, S. (1986). *In the Patient's Best Interest*. New Brunswick, NJ: Rutgers University Press.

Gahan, C., & Hannibal, M. (1998). *Doing Qualitative Research Using OSR NUDIST*. London: Sage.

Gollub, E. L (1999). Human Rights Is a U.S. Problem, Too: The Case of Women and HIV. *American Journal of Public Health, 89*, 1479–1482.

Green, R. (1990). American Indian Women: Diverse Leadership for Social Change. In: L. Albrecht & R. M. Brewer (Eds), *Bridges of Power: Women's Multicultural Alliances*. Philadelphia: New Society.

Grinstead, O. A (1997). HIV Counseling for Behavior Change. *AIDS Education and Prevention, 9*, 125–132.

Heise, L. L., & Elias, C. (1995). Transforming AIDS Prevention to Meet Women's Needs: A Focus on Developing Countries. *Social Science and Medicine, 40*, 931–943.

Hirano, D., Slaughter, F. R., Sands, L., & Gellert, G. A. (1994). Increasing the Effectiveness of HIV Counseling. *American Journal of Public Health, 84*, 1179–1180.

Hudson, A. L., & Morris, R. I. (1994). Perceptions of Social Support of African Americans with Acquired Immunodeficiency Syndrome. *Journal of National Black Nurses Association, 7*, 36–49.

Hunter, N. D (1994). Complications of Gender: Women, AIDS, and the Law. In: B. E. Schneider & N. E. Stoller (Eds), *Women Resisting AIDS* (pp. 32–56). Philadelphia: Temple University Press.

Ickovics, J. R., Morrill, A. C., Beren, S. E., Walsh, U., & Rodin, J. (1994). Limited Effects of HIV Counseling and Testing for Women. *Journal of the American Medical Association, 272*, 443–448.

Illinois (1997). *Illinois Compiled Statutes, Public Health, 410 ILCS 305/3–7.* Retrieved October 1999 (http://legis.state.il.us/ilcs/ch410/ch410act305.htm)

Kamb, M. L., Fishbein, M., Douglas, J. M., Rhodes, F., Rogers, J., Bolan, G., Zenilman, J., Hoxworth, T., Malotte, K., Iatesta, M., Kent, C., Lentz, A., Graziano, S., Byers, R. H., & Peterman, T. A. (1998). Efficacy of Risk-Reduction Counseling to Prevent Human Immunodeficiency Virus and Sexually Transmitted Diseases. *Journal of the American Medical Association, 280*, 1661–1667.

Landis, S. E., Earp, J. L., & Koch, G. G. (1992). Impact of HIV Testing and Counseling on Subsequent Sexual Behavior. *AIDS Education and Prevention, 4*, 61–70.

Lather, P., & Smithies, C. (1997). *Troubling the Angels.* Oxford: Westview Press.

Markowitz, M., Saag, M., Powderly, W. G., Hurley, A. M., Hsu, A., Valdes, J. M., Henry, D., Sattler, F., La Marca, A., Leonard, J. M., & Ho, D. D. (1995). A Preliminary Study of Ritonavir, an Inhibitor of HIV-1 Protease, to Treat HIV-1 Infection. *The New England Journal of Medicine, 333*, 1534–1539.

Mishler, E. G (1984). *The Discourse of Medicine: Dialectics of Medical Interviews.* Norwood, NJ: Ablex.

Murphy, J., Mueller, G., & Whitman, S. (1996). Epidemiology of AIDS Among Hispanics in Chicago. *Journal of Acquired Immune Deficiency Syndromes and Human Retrovirology, 11*, 83–87.

New York (1988). *New York State Consolidated Laws: Public Health Article 27-F Section 2781, 1–5.* Retrieved October 1999 (http://caselaw.findlaw.com/scripts/getcode.pl?frame = right2&type = srchres&)

Niemcryk, S. J., Bedros, A., Marconi, K. M., & O'Neill, J. F. (1998). Consistency in Maintaining Contact with HIV-Related Service Providers: An Analysis of the AIDS Cost and Services Utilization Study. *Journal of Community Health, 23*, 137–152.

Osmond, M. W., Wambach, K. G., Harrison, D. F., Byers, J., Levine, P., Imershein, A., Quadagno, D. M. (1993). The Multiple Jeopardy of Race, Class, and Gender for AIDS Risk Among Women. *Gender & Society, 7*, 99–120.

Otten, Mac W., Zaidi, A. A., Wroten, J. E., Witte, J. J., & Peterman, T. A. (1993). Changes in Sexually Transmitted Disease Rates after HIV Testing and Posttest Counseling, Miami, 1988 to 1989. *American Journal of Public Health, 83*, 529–533.

Phillips, K. A., Morrison, K. R., Sonnad, S. S., & Bleecker, T. (1997). HIV Counseling and Testing of Pregnant Women and Women of Childbearing Age by Primary Care Providers: Self-Reported Beliefs and Practices. *Journal of Acquired Immune Deficiency Syndromes and Human Retrovirology, 14*, 174–178.

Quinn, S. C. (1993). Perspective AIDS and the African American Woman: The Triple Burden of Race, Class, and Gender. *Health Education Quarterly, 20*, 305–320.

Russell, J. M., & Smith, K. (1998). HIV Infected Women and Women's Services. *Health Care for Women International, 19*, 131–139.

Schneider, B. E (1992). AIDS and Class, Gender, and Race Relations. In: J. Huber & B. E. Schneider (Eds), *The Social Context of AIDS* (pp. 19–43). London: Sage.

Sikkema, K. J., & Bissett, R. T. (1997). Concepts, Goals, and Techniques of Counseling: Review and Implications for HIV Counseling and Testing. *AIDS Education and Prevention, 9*(B), 14–26.

Smith, D. E (1987). *The Everyday World As Problematic: A Feminist Sociology.* Boston: Northeastern University Press.

Sobo, E. J (1995). HIV Testing and Wishful Thinking. In: E. J. Sobo (Ed.), *Choosing Unsafe Sex: AIDS-Risk Denial Among Disadvantaged Women* (pp. 140–156). Philadelphia: University of Pennsylvania Press.

Strauss, A. L. & Corbin, J. (1998). *Basics of qualitative research: Techniques and procedures for developing grounded theory.* Thousand Oaks, CA: Sage.

Triechler, P. A. (1988). AIDS, Gender, and Biomedical Discourse: Current Contests for Meaning. In: E. Fee & D. M. Fox (Eds), *The Burdens of History* (pp. 190–266). Berkeley: University of California Press.

U.S. Census Bureau (2000). *Poverty 1999.* Retrieved May 8, 2000 (http://www.census.gov/hhes/poverty/threshld/thresh99.html)

Waitzkin, H. (1989). A Critical Theory of Medical Discourse: Ideology, Social Control, and the Processing Social Context in Medical Encounters. *Journal of Health and Social Behavior, 30*, 220–239.

Weinhardt, L. S., Carey, M. P., Johnson, B. T., & Bickham, N. L. (1999). Effects of HIV Counseling and Testing on Sexual Risk Behavior: A Meta-Analytic Review of Published Research, 1985–1997. *American Journal of Public Health, 89*, 1397–1405.

Weissman, G., Melchior, L., Huba, G., Altice, F., Booth, R., Cottler, L., Genser, S., Jones, A., McCarthy, S., Needle, R., & Smereck, G. (1995). Women Living with Substance Abuse and HIV Disease: Medical Care Access Issues. *Journal of the American Medical Women's Association, 50*, 115–120.

West, C. (1984). *Routine Complications: Troubles With Talk Between Doctors and Patients.* Bloomington, IN: Indiana University Press.

West, C. (1993). Reconceptualizing Gender in Physician-Patient Relationships. *Social Science and Medicine, 36*, 57–66.

Wolitski, R. J., MacGowan, R. J., Higgins, D. L., & Jorgensen, C. M. (1997). The Effects of HIV Counseling and Testing on Risk-Related Practices and Help-Seeking Behavior. *AIDS Education and Prevention, 9* (Supplement B), 52–67.

BREAST CANCER: AN EXAMINATION OF THE SOCIAL CORRELATES RELATIONSHIP TO DIAGNOSIS AND TREATMENT

Carla D. Edwards, Amber McClave and Yvonne J. Combs

ABSTRACT

This chapter examines the existing research on breast cancer screening, diagnosis, and treatment. The authors identify social and demographic factors which may explain the decision making process of women diagnosed with breast cancer. Based on an examination of the existing research on breast cancer diagnosis and treatment, the authors have concluded that social and demographic factors, such as race, culture and ethnicity, age, socioeconomic status, social support, and quality of life, have varying influence on the decisions made by women diagnosed with breast cancer.

INTRODUCTION

According to the Center for Disease Control and Prevention's 1997 death statistics, cancer is the leading cause of death for women between the ages of twenty-five and forty-four (Hoyert, Kochanek & Murphy, 1999). Breast cancer is the most common form of cancer among women in the United States,

Health, Illness, and Use of Care: The Impact of Social Factors, Volume 18, pages 35–51.

ISBN: 0-7623-0740-4

striking approximately 182,000 women and killing approximately 46,000 women each year (Carlson, Eisenstat & Ziporyn, 1996). It has become a major health concern among women in the United States and abroad (Carlson et al., 1996; Canada News Wire, 1997; Wilkinson & Kitzinger, 1994). Breast cancer has been pronounced as one of the most feared diseases by American women accompanied with the distress and fear of the traditional treatment, a mastectomy or the loss of a breast (Carlson et al., 1996; Wilkinson & Kitzinger, 1994). This fear has been attributed to the major preoccupation with sexuality and body image fueled by male dominance in the majority of societies. According to Wilkinson and Kitzinger (1994), the preoccupation with body image typically overrides a woman's concern with the functional aspects of breast, such as breast feeding. Even though, breast cancer is a disease affecting women, much of the research on breast cancer and issues related to breast cancer is a reflection of the dominance of men in the medical profession (Wilkinson & Kitzinger, 1994).

Much of the non-clinical research on breast cancer has focused on the issue of non-compliance with breast cancer screening procedures among women of various ages (Caplan, Wells & Haynes, 1992; Champion, 1992; Crane, Kaplan, Bastani & Scrimshaw, 1996; Webber, Fox, Zhang & Pond, 1996), from different regions with different levels of access (Hurley, Huggins, Jolley & Reading, 1994; Schoen, Marcus & Braham, 1994), and from particular racial, ethnic, and socioeconomic backgrounds (Crane et al., 1996; Greenwald, Kramer & Weed, 1995; Hurley et al., 1994; Jacob, Spieth & Penn, 1993; Webber et al., 1996). Researchers have focused on the psycho-social factors that influence the use of different breast cancer screening techniques and the level of compliance by women introduced to various techniques (Andersen, 1986; Brand & van Keep, 1978; Haber, 1995; Zapka, Harris, Stoddard & Costanza, 1991). Additionally, social-psychological scientists have researched how diagnosis, treatment, and survivorship is influenced by coping mechanisms (Ell, Nishimoto, Mantel & Hamovitch, 1992; Gilber, 1997; Heuser, 1991; Roberts, Cox, Shannon & Wells, 1995) and the level of economic (Barofsky, 1989) and social support available to the patient and her family (Ariel & Cleary, 1987; Baum, Saunders & Meredith, 1994; Kameg, 1995; Satariano & DeLorenze, 1996; Virag, 1995; Weinberg, Uken, Schmale & Adamek, 1995). Much of the clinical research on breast cancer focuses on identifying causative agents for the purposes of finding a cure and/or improving treatment options (Fentiman, 1990; "FDA OKs", 1998; Greenwald et al., 1995; Batholomew & Craig, 1984; Pelton & Overholser, 1994; Venet, 1979).

To date, there is very little research on the decision making process employed by women once they have been diagnosed with breast cancer

(Andersen, 1993; Lauver, 1994; Lauver, Coyle & Panchmatia, 1995). The research has focused on the decision making process of the woman's doctor (Fentiman, 1990; McGuire, Tandon, Allred, Chamness, Ravdin & Clark, 1992) and the woman's behavior regarding breast cancer screening and treatment (Jacob et al., 1993; Crane et al., 1996; Greenwald et al., 1995; Hurley et al., 1994; Schoen et al., 1994; Webber et al., 1996), but it has not taken into consideration emotional and psychological trauma or the burdens women face in their everyday lives (Kasper, 1994; Wilkinson & Kitzinger, 1994). The existing research on women and breast cancer presumes that the decision making power is in the hands of the doctor and that the role of the woman is to comply with the assigned treatment. The majority of research on breast cancer fails to address the reasons why certain treatment decisions are made once a woman has been diagnosed. This position study examines the social factors which may influence a women's treatment decision, assuming that she has a right to actively participate in the process of making decisions not only about breast cancer treatment, but about her own morbidity and mortality as well.

More female centered research is needed to determine the factors that influence a woman's decision regarding breast cancer treatment. Naturally, the diagnosis and prognosis will have a profound effect on the available treatment options and the decisions made by the doctor, but how does a woman's age, race or culture, socioeconomic status, level of social, emotional, and economic support, and concerns about the quality of her life influence her treatment decision? In this chapter, we explore these social factors and we propose that future research investigate the direct and/or indirect effects of these factors on the decision making process employed by women faced with breast cancer, one of the deadliest and most feared diseases known to women today.

According to a national survey of Canadian women newly diagnosed with breast cancer, women want to know all of their treatment options upon diagnosis and they want to be a part of the treatment decision making process. When asked about the role of the patient in making decisions about treatment or therapy, 77% of the women preferred to be a part of the decision-making process, if not make the decision themselves. Only 22% preferred that their doctors make the decision for them (Canada News Wire, 1997). This study indicates that women do not prefer to be passive adherents to a treatment regime that can effect the course and quality of their lives. However, many women reported that they were not given sufficient information about alternative treatments, funding for treatment, and the potential side effects of the prescribed treatment upon diagnosis (Canada News Wire, 1997). The invasiveness, the cost, and the effects of the treatment are important factors to consider when making a treatment decision, yet women are hampered from participating in the decision making process because this information is not always available to them at the time of diagnosis.

FACTORS RELATED TO SCREENING AND DIAGNOSIS

There is a considerable amount of research on the factors related to breast cancer screening and diagnosis among women from various backgrounds. By reviewing the literature on screening and diagnosis, we have been able to identify some of the factors which may influence treatment decisions among women diagnosed with breast cancer. Researchers have found that gender and age of the physician (Andersen & Urban, 1997; Schoen et al., 1994), the patient's race (Caplan et al., 1992; Jacob et al., 1993), age (Caplan et al., 1992; Champion, 1992; Crane et al., 1996; Webber et al., 1996), and socioeconomic status (Crane et al., 1996; Hurley et al., 1994; Schoen et al., 1994) are significantly related to the use of breast cancer screening. Andersen and Urban (1997) found gender of the physician to be the most significant variable related to breast cancer screening among patients with similar sociodemographic and psychosocial backgrounds. Their research found that women with female physicians were almost twice as likely to receive a mammogram than women with male physicians. Additionally, Schoen and associates (1994) found that first-year postgraduate physicians were more likely to screen their patients than the more senior physicians on staff. Hence, the findings of Anderson and Urban (1997) and Schoen et al. (1994) demonstrate that their are sociodemographic factors influencing the behavior of the physicians which in turn could have an impact on the future decision making power of the patient when diagnosed with cancer.

Furthermore, the patient's race, age, and socioeconomic status have been reported as having a significant effect on decisions about and participation in breast cancer screening programs. Older minority women receive less clinical examinations than white women (Caplan et al., 1992) and African-American women are less likely to perform breast self examinations due to various factors (Jacob et al., 1993). A recent study of breast-self examinations (BSE) among African-American women reported that African-American women were less likely to perform BSE than white women and among the African-American women who performed BSE very few performed the procedure correctly (Jacob et al., 1993). Similar to research with the majority population, Jacob and her associates (1993) found that among the African-American women in their study, older women with high degrees of confidence, from a high socio-economic status, and who had regular contact with health professionals were more likely to perform BSE and to perform it correctly. Therefore, it is questionable as to whether race is as significant a factor as age, self-efficacy, and socioeconomic status in the use of BSE.

Webber and colleagues (1996) found that age and the results of the patients' initial screening were significantly related to rates of follow-up screening among women participating in the Breast and Cervical Cancer Control Program in California. Age was the only demographic variable to retain its significance in the analysis. Unlike Jacob et al.'s (1993) study, older women in this program were less likely to follow-up after receiving an initial abnormal breast screen. Champion (1992) conducted a study on BSE among women sixty-five and older and she found that health beliefs and attitudes were significantly related to the use of BSE among older women. Women who believed that they were at risk, that the illness was serious, and that their behavior would have an effect on the prevention or amelioration of the disease were more likely to comply with the recommended BSE protocol. These beliefs are indicative of the health belief model which has shown to be significantly related to health seeking behavior among a variety of populations (Katz, 1993).

Additional research has confirmed that women with higher levels of education, fewer barriers to health resources, and higher socioeconomic statuses are more likely to comply with breast cancer screening recommendations (Crane et al., 1996; Hurley et al., 1994; Schoen et al., 1994). A study of Belgian women found that participation in screening was highest among the middle class. The researchers found that women who were socially integrated and who maintained a belief in their ability to control health and illness were more likely to participate in screening programs (Brand & van Keep, 1978).

These results indicate that there is a not only a relationship between the status and behavior of the health care provider and the sociodemographic background of the patient, but beliefs and attitudes about health seeking behavior are related to diagnosis and screening as well. These sociodemographic and psychosocial factors are missing from consideration when looking at the treatment decisions of women. Researchers have assumed that treatment decisions are the responsibility of the physician and that the decision is made using prognostic factors and simply weighing the medical benefits against the medical risks (McGuire et al., 1992). The research on the various factors influencing screening and diagnosis demonstrates that health seeking behavior is influenced by a variety of factors and not all of the factors are objective, rational, and easily observable as orthodox medicine and traditional science presumes.

FACTORS RELATED TO TREATMENT

Treatment Options And Decisions

There are over fourteen different types of breast cancer treatments (Strax, 1974). The available treatments range from changes in lifestyle and nutritional habits to hormone therapy to radiation or chemotherapy to surgical removal of the cancer and/or the entire breast. The non-invasive therapies are considered alternative therapies, while the surgical procedures are classified as conservative or aggressive depending on how much of the breast and chest cavity is removed (Fentiman, 1990; Pelton & Overholser, 1994; Venet, 1979). The particular combination of these treatments that is right for one type of cancer or one woman may be completely inappropriate for another. Among the many factors to consider when deciding on the appropriate treatment are the size, nature, and stage of the cancer; the woman's menopausal status, age, and overall health; and the potential side effects, cost, convenience, and time commitment entailed by the treatment (Carlson et al., 1996).

How do women decide what to do when faced with various treatment options? Due to the social variations found in the research on breast cancer screening and diagnosis and the variations in the incidence and prevalence of the disease among women from different sociodemographic backgrounds, we hypothesize that there are probably similar variations in the treatment decision making process. How women choose to deal with breast cancer may be influenced by their age, race, socioeconomic status, social support networks, and concern with the quality of their lives.

When a woman is diagnosed with breast cancer, often she is not given adequate information despite the fact that she is about to make a decision that will affect every aspect of her life. Her first decision involves how much tissue to remove – just the lump itself or the entire breast. This decision leads to other decisions about radiation therapy, reconstructive surgery, chemotherapy, and/or hormonal therapy.

Even though it is clear that the most important factor to consider in the decision making process is the stage of cancer at the time of diagnosis and prognosis (Strax, 1974), we propose that there are social factors that may have a direct or indirect effect on a woman's treatment decisions. Hypothetically, these social factors may account for the variations in treatment decisions thus influencing morbidity and mortality rates of women diagnosed with breast cancer. Timing of the diagnosis is crucial when trying to fight breast cancer. But unfortunately, not all women are screened and diagnosed within the necessary time frame to improve chances of survival. Certain populations of women are diagnosed earlier than other populations of women, and therefore, have a better survival rate. The fight against breast cancer actually begins with the ability to obtain the correct knowledge and information and access to the required diagnostic tests and medical treatment. These factors will affect a woman's decision making at the time of diagnosis. If she is knowledgeable about what necessary precautions to take and is able to have regular checkups, the cancer will be detected earlier, altering the range of her treatment options. Yet, if a woman is not diagnosed until she is in the late stages of the disease, her decision making power becomes very limited.

Age

Breast cancer rates began rising around 1940, but increased sharply around 1980. Most of the increase in the United States has been among women over age fifty, even though it is younger women who seem to fear breast cancer the most (Carlson et al., 1996). Only 5% of breast cancers occur in women under forty, and 25% occur in women under the age of fifty. Breast cancer is one of the most serious health problems for women over the age of sixty-five (Guillory, 1994; Mills & Blank, 1992). A major reason why death rates from breast cancer have risen in older women is that mortality from other causes has declined. Since women are living longer in general, the number surviving long enough to develop breast cancer has risen (Carlson et al., 1996). One of the problems with these changing demographics is that elderly women are not targeted for early detection (Guillory, 1994). For women over fifty, breast cancer screening and early treatment are the most effective weapons in fighting the disease (Mills & Blank, 1992).

Another problem associated with breast cancer among elderly women is the extensive use of aggressive therapies that can hamper the quality of life of aged women. Our cultural notions of beauty are that once a woman is past a certain age, she is no longer sexually attractive or active, therefore, her breasts are no longer of use. This male dominated sense of beauty and the late stage of diagnosis have both contributed to a disproportionate rate of radical mastectomies among elderly women with breast cancer. Further investigation may reveal that our cultural notions of age and beauty are influencing treatment decisions and ultimately effecting the lives of older women with breast cancer. Are doctors more likely to push for more invasive surgery when the woman is past a certain age, and if so, is this due to a greater risk from biological age or to preconceived notions about elderly women and sexuality? This cannot be answered by the current research.

Race, Ethnicity, and Culture

It is very difficult to distinguish between the use of the terms race, culture, and ethnicity in research literature. Race may refer to people with the same skin color or to the genetically or socially inherited factors associated with a group of people descending from the same evolutionary lineage. Ethnic group has been defined as a group of people distinguished by cultural and national characteristics, but on some occasions the word has been alternated with race. Additionally, the term culture or cultural group has been interchanged for the words race and ethnicity when serving a particular political purpose. In this study, we will use the terms interchangeably and loosely to identify characteristics of particular groups in the United States that have been classified as minorities. Perhaps, a woman's minority status, racial or ethnic background or cultural beliefs will influence the decisions made about her treatment for breast cancer.

The rate of survival for cancer patients has improved over the past thirty years, but the rate of survival is still significantly less for African-American women than for white women (Stellman, 1987), even though the incidence of breast cancer is higher among whites than any other racial or ethnic group (Greenwald et al., 1995). According to Greenwald and associates (1995), white women have the highest rate of cancer (92.9 per 100,000). The rate for African-Americans is 75.2. It is 57.8 for Chinese Americans, 55.0 for Japanese Americans, 52.1 for Hispanic Americans, and 21.3 for Native Americans. The differences between the survival rates of Whites, African-Americans, Hispanic-Americans, and other minority groups living in the United States have been attributed to age differentials, genetic susceptibility, migratory and cultural differences, socioeconomic status, susceptibility to infectious diseases, and deficiencies in data collection and analysis (Greenwald et al., 1995). Minorities experience poverty, cross-cultural related stress, environmental hazards, and discrimination at a disproportionate rate when compared to whites resulting in increased risk of disease and higher morbidity and mortality rates at younger ages. However, some research has shown that even though a woman's racial classification can place her at a greater risk of poor health outcomes, there are cultural norms and values which may protect some minority women from disease.

A strong emphasis on family and intergenerational relationships characterizes most minority communities, but the strong sense of family and community has proven to be a significant health protective factor for Hispanic-Americans (Aragon de Valdez & Gallegos, 1982; Suarez, 1994; Stellman, 1987). The rates of breast cancer are lower among Hispanic women than white or African-

American women and they are even lower among foreign born Hispanic women than those born within the United States. These findings suggest that there are potential cultural and environmental factors contributing to the different rates of cancer among women from different racial and ethnic backgrounds (Stellman, 1987). According to a study by the Council on Scientific Affairs (1991), Hispanics have a lower incidence of breast cancer. Their high fertility rate and the young age at which they give birth are protective factors against breast cancer. Given the value placed on family by many Hispanic-Americans, one may assume that this cultural value would not only influence the prevalence of breast cancer among Hispanic women, but it may also influence the decisions made by Hispanic women who are diagnosed with cancer. The role of the family would become a significant factor in the treatment decision making process, because the potential for the family to be effected is so great.

Similarly, cultural notions of beauty, nutritional habits, and religious practices or spiritual beliefs may influence the decisions made by women from different backgrounds. To date there is no or very limited published research on how these indicators of race, culture, and ethnicity influence the treatment decisions made by women. But given the significant differences between the morbidity and mortality rates of White and minority women with breast cancer, it is plausible that these factors play a huge role in health seeking behavior.

Socioeconomic Status

Generally, socioeconomic status (SES) refers to the financial, educational, and social status of an individual or group of people. Different researchers define it in different ways, but for the purposes of this study, the general definition is sufficient. As noted earlier, SES is significantly related to screening and diagnosis of breast cancer and it is significantly related to breast cancer survival rates (Stellman, 1987). Women from lower SES have lower rates of survival than women from higher SES. The lower rates of survival have been attributed to barriers to early detection (see section on Screening and Diagnosis for references), availability of health insurance (Hadley & Mitchell, 1997), access to routine health care, lower levels of education and lower incomes (Stellman, 1987), and lifestyle or behavioral factors (Freimuth & Mettger, 1990; Katz, 1993). Researchers have found that early detection was the most significant predictor of five-year survival rates for women effected with breast cancer. But the differences in who received and did not receive screening varied considerably by SES (Stellman, 1987).

Related breast cancer survival research has focused on the relationship between access to care and early detection, recovery and survival rates (Mills & Blank, 1992). Hadley and Mitchell (1997) actually conducted a study on the relationship between breast cancer treatment decisions and type of health care coverage. Their research found that women enrolled in Health Maintenance Organizations (HMOs) were less likely to receive breast conservation therapy or surgery than women with other types of private insurance. The use of mastectomies was higher and the length of stay in the hospital after surgery was shorter for women with HMOs. Due to the limitations of this study (including their inability to control for other factors, such as stage of disease and the additional SES factors) it is not advisable to make generalizations or conclusions about the relationship between decision making and insurance coverage using these findings. But Hadley and Mitchell's study (1997) helps justify the need to do further research on the factors influencing a women's treatment decision. This research suggest that a relationship exists between health care coverage, a socioeconomic variable and treatment decisions.

Other studies have found that uninsured women are more likely to have higher stages of the disease at diagnosis and are less likely to receive aggressive treatment. Weiss (1997) reports that "after adjusting for age, race, stage of diagnosis, and household income, the researchers found that uninsured women and those on Medicare were between 40 and 89% more likely to die in the 7.5 years after initial diagnosis than those privately insured" (31). This research suggests that SES, as defined by access to adequate health insurance, has an effect on screening and diagnosis, type of treatment received and ultimately, a woman's chance of survival. Almost one-quarter of Americans are without health insurance for at least one month in a two year period and approximately 15% of Americans are without health insurance at any given point in time (Weiss, 1997).

With the increasing number of women living in poverty, the low levels of health insurance is a particular concern for low-income, unmarried, minority women and mothers. Lack of health care coverage and low SES is a barrier to care and a huge obstacle to bear when faced with a life-threatening illness like breast cancer. First, the woman must have the cognitive and language skills to understand her treatment options, then she must weigh her ability to pay for the most effective treatment while battling the invisible discriminatory decisions made by the health care provider. In most cases, the health care provider controls the woman's access to information about her health status, treatment options and risks. By taking a closer look at the relationship between SES and breast cancer treatment decision making, researchers can provide health educators and service providers with information to ameliorate some of the problems.

Support Networks

Neuling and Winefield (1988) studied social support and recovery after surgery for breast cancer. Their results showed that frequency of support from all sources decreased as time from surgery passed, while satisfaction with support varied with the type of support given and the source from which it was received. Satisfaction with social support was matched with measures of adjustment, and it was found that those satisfied with support from family members were significantly less anxious and depressed while in the hospital than were those who were not satisfied with support from this source. In addition, one month after the operation, anxiety and depression levels were significantly related to satisfaction with support from surgeons; and after three months, anxiety and depression measures were significantly related to satisfaction with support from both family members and surgeons. This is an important finding as the typical measure of social support does not include support from the physician.

Peters-Golden (1982) researched varied perceptions of social support among people with and without cancer. The non-cancer patients were more likely to report that a stigma and avoidance behavior is typically attributed to cancer patients, yet the non-cancer patients believed that they themselves would have a dense network of support to call upon if stricken with cancer. In other words, non-cancer patients believe that cancer patients are avoided by others due to their illness, but they believe that if the tables were turned (and they were stricken with cancer) they would not avoid others and that others would not avoid them. Also, patients assert that the support extended to them is often inappropriate in nature, due to the "mistaken assumption on the part of health individuals that the most salient concern of a mastectomized woman is the loss of her breast, rather than the fact that she has cancer."

Researchers have found that, in addition to stage of disease, the number of supportive friends, number of supportive persons, whether the woman worked, whether she was unmarried, the extent of contact with friends and the size of her social network are all factors that significantly and independently effect a woman's chance of survival (Waxler-Morrison, Hislop & Kan, 1991). Thus, the social context is invariably important to survival. In addition, Neale, Tilley and Vernon (1986) found that marital status had an independent and statistically significant effect on differences in survival even when age, socioeconomic status, delay in treatment, and stage of disease were included in the model.

Clearly, then, there is a relationship between perceived and available social support, recovery, healing, and survival. Studies have shown that high levels of support, whether it is a computer mediated support group, a spouse, family member, friend, or rehabilitative service, increases the patient's ability to cope and to re-establish a normal lifestyle (Gilber, 1997; Kameg, 1995; Satariano & DeLorenze, 1996; Weinberg, Uken, Schmale & Adamek, 1995). The mere notion that a woman has people that are close to her to depend on, including a positive an open relationship with her medical expert, allows her a better chance of survival. We have no data to show that the same perception of social support affects what decisions women make regarding treatment, but is it not logical to assume that such a relationship exists? A testable hypothesis is, women with husbands, parents, children, and friends, who are ready and willing to take care of them, their homes, their businesses, and their children, after surgery or during chemotherapy are more likely to undergo such treatment, than women with no one to support them. Such practical concerns are lacking as contributing factors to treatment variations in the literature.

Quality of Life

From a feminist perspective quality of life after breast cancer surgery has very little to do with a woman's ability to sexually attract or satisfy her mate, but most research on quality of life after breast cancer mentions concerns about sexuality, body-image and self-concepts (Andersen, 1986; Ghizzani, Pirtoli, Bellezza & Velicogna, 1995; Kaplan, 1992; Reaby, Hort & Vandevord, 1994; Stellman, 1987; Venet, 1979; Ariel & Cleary, 1987). Women have reported concerns over the amount of financial and emotional support available to them, their ability to manage daily activities, their ability to return to work, and anxiety and depression, but very few have listed sexual functioning as a primary concern (Haber, 1995).

The majority of research indicates that post-mastectomy women suffer with various levels of disability, body-image dissatisfaction and poor feminine self-image. However, researchers have found that the quality of life experienced by women who have had breast cancer surgery was not dependent on their attitudes toward their appearance, but it was significantly related to the amount of social support available to the patient, the amount of stress the individual experienced, the existence of other chronic diseases, and the amount of control the patient felt she had in the decisions affecting her life (Stellman, 1987).

Street and Voigt (1997) conducted a study on the relationship between the amount of control the patient believed she had over treatment decisions, the patient's participation in the consultation with her doctor about treatment, and the patient's reported quality of life six months to one year after breast cancer treatment. It was found that quality of life was related to the patient's level of involvement during the consultation with the doctor and with the patient's feeling of responsibility for her treatment based on having had options to choose from. Street and Voigt (1997) reported that those women who felt that they were responsible for their treatment decision and who discussed their options with their doctor reported having a better quality of life post surgery than other women. This study shows that when women are involved in treatment decisions their quality of life improves, but does a woman's concern about her future quality of life drive her decision? Future research should examine this question. Researchers have reported that women experience physical and emotional distress after breast cancer surgery (Haber, 1995; Venet, 1979) and a few studies have indicated that the fear and stress associated with diagnosis can be severe as well (Baum et al., 1994; Ariel & Cleary, 1987). Therefore, we hypothesize that the concerns generated by knowing other women who have suffered adversely after surgery and whose quality of life has deteriorated may influence the decisions about treatment by women who are newly diagnosed.

CONCLUSION

According to *The Harvard Guide to Women's Health*, the gradual increase in breast cancer from 1940 to 1980 has been attributed to social and environmental factors, such as lifestyle changes, improved nutrition (which increases a woman's lifetime exposure to estrogen, by allowing menarche to start earlier), delayed childbearing, and possible exposure to environmental toxins. But the steep rise in the number of cases of cancer diagnosed since 1980 has been attributed largely to breast cancer screening and early diagnosis. Since its inception, mammography, one of the many screening techniques, has picked up cases of breast cancer that previously would have advanced to an untreatable stage when finally diagnosed. Between 1989 and 1992 death rates from breast cancer in American women fell five percent, the largest short-term decline since 1950. For women in their thirties, forties, and fifties mortality has decreased as much as nine percent (Carlson et al., 1996). But the death rate for breast cancer among African American and other minority women rose during this period and it continues to rise. In this chapter, we have discussed the reasons for the differences between screening practices, diagnosis, treatment, and recovery among women of different ages, racial and ethnic backgrounds, and socioeconomic statuses.

Our knowledge of treatment decisions is largely based upon the stage of the disease at time of diagnosis. It is clear that the development of the cancer is directly related to treatment options typically offered to the patient. This contributes to different outcomes in treatment and survival rates for individual women. Polissar and Finley (1985) found that use of the biopsy-mastectomy procedure was associated with younger age of the patient and less suspicious symptoms or mammogram results while older women were given more invasive treatments. Other studies have shown us that poorer and less educated women receive fewer options than socially integrated and upperclass women. Additionally, women of different racial and ethnic backgrounds have varying cultural beliefs which influence their health-seeking behavior and in turn may influence their decisions about breast cancer treatment. Understandably, the urging of health care professionals for early diagnosis through mammography and breast self-examination is motivated by their desire to reduce the differences in treatment and outcomes. However, the stage of the disease at diagnosis is not the only factor which contributes to health seeking behavior. The studies we have cited have driven us to conclude that sociodemographic factors such as race, culture and ethnicity, age, socioeconomic status, social support, and quality of life are variables worth considering when exploring the treatment decisions among women diagnosed with breast cancer.

REFERENCES

Andersen, B. L. (Ed.) (1986). *Women with Cancer: Psychological Perspectives*. New York: Springer-Verlag.

Andersen, G. (1993). *50 Essential Things to Do When the Doctor Says It's Cancer*. New York: Penguin Group.

Andersen, M. R., & Urban, N. (1997). Physician Gender and Screening: Do Patient Differences Account for Differences in Mammography Use? *Women and Health, 26*, 29–39.

Aragon de Valdez, T., & Gallegos, J. (1982). The Chicano Familia in Social Work. In: J. W. Green (Ed.), *Cultural Awareness in the Human Services*. Englewood Cliffs, NJ: Prentice-Hall.

Ariel, I. M., & Cleary, J. (1987). *Breast Cancer: Diagnosis and Treatment*. New York: McGraw-Hill Book Company.

Barofsky, I. (1989). *Work and Illness: The Cancer Patient*. New York: Praeger Publishers.

Batholomew, J., & Craig, N. J. (1984). *Environment and Health in Louisiana: The Cancer Problem, Executive Summary. Report of Task Force on Environmental Health*. Baton Rouge, LA: Louisiana State Planning Office.

Baum, M., Saunders, C., & Meredith, S. (1994). *Breast Cancer: A Guide for Every Woman*. Oxford: Oxford University Press.

Brand, P. C., & van Keep, P. A. (Eds) (1978). *Breast Cancer: Psycho-social Aspects of Early Detection and Treatment*. Baltimore, MD: University Park Press.

Caplan, L. S., Wells, B. L., & Haynes, S. (1992). Breast Cancer Screening among Older Racial/ethnic Minorities and Whites: Barriers to Early Detection. *Journals of Gerontology, 47*, 101–106.

Carlson, K. J., Eisenstat, S. A., & Ziporyn, T. (1996). *The Harvard Guide to Women's Health*. Boston, MA: Harvard University Press.

Canada News Wire (CNW) (1997). National Study of Breast Cancer Patients Reveals How Women Feel about Information Received During Treatment. *www.newswire.ca*

Champion, Victoria (1992). Breast Self-examination in Women 65 and Older. *Journals of Gerontology, 47*, 75–80.

Council on Scientific Affairs (1991). Hispanic Health in the United States. *JAMA, 265*, 248–252.

Crane, L. A., Kaplan, C. P., Bastani, R., & Scrimshaw, S. C. M. (1996). Determinants of Adherence among Health Department Patients Referred for a Mammogram. *Women and Health, 24*, 43–64.

Ell, K., Nishimoto, R., Mantel, J., & Hamovitch, M. (1992). Coping with Cancer: a Comparison of Older and Younger Patients. *Journal of Gerontological Social Work, 19*, 3–27.

FDA OKs new breast cancer drug (1998). *The Gainesville Sun.*

Fentiman, I. S (1990). *Detection and Treatment of Early Breast Cancer.* Philadelphia: J. B. Lippincott Company.

Freimuth, V. S., & Mettger, W. (1990). Is There a Hard-to-reach Audience? *Public Health Reports, 105*, 232–238.

Ghizzani, A., Pirtoli, L., Bellezza, A., & Velicogna, F. (1995). The Evaluation of Some Factors Influencing the Sexual Life of Women Affected by Breast Cancer. *Journal of Sex and Marital Therapy, 21*, 57–63.

Gilber, O. (1997). Cancer Caregiver Support Group: a Model for Intervention. *Clinical Gerontologist, 18*, 31–37.

Greenwald, P., Kramer, B. S., & Weed, D. L. (1995). *Cancer Prevention and Control.* New York: Marcel Dekker, Inc.

Guillory, J. A (1994). Breast Cancer: a Serious Threat to Elderly Women. *Journal of Women and Aging, 6*, 151–164.

Haber, S. (Ed.) (1995). *Breast Cancer: A Psychological Treatment Manual.* New York: Springer Publishing Company.

Hadley, J., & Mitchell, J. M. (1997/98). Breast Cancer Treatment Choice and Mastectomy Length of Stay: a Comparison of HMO and Other Privately Insured Women. *Inquiry, 34*, 288–301.

Heuser, L. (1991). Perceptions of Women's Emotional Reactions to Breast Cancer. *Sociology and Social Research, 75*, 219–226.

Hoyert, D. L, Kochanek, K. D., & Murphy, S. L. (1999). *Deaths: Final Data for 1997. (National Vital Statistics Reports Volume 47, Number 19).* National Center for Health Statistics, Center for Disease Control and Prevention, U.S. Department of Health and Human Services.

Hurley, S. F., Huggins, R. M., Jolley, D. J., & Reading, D. (1994). Recruitment Activities and Sociodemographic Factors That Predict Attendance at a Mammographic Screening Program. *The American Journal of Public Health, 84*, 1655–1959.

Jacob, T. C., Spieth, L. E., & Penn, N. E. (1993). Breast Cancer, Breast Self-examination, and African-American Women. In: B. Bair & S. E. Cayleff (Eds), *Wings of Gauze: Women of Color and the Experience of Health and Illness.* Detroit: Wayne State University Press.

Kameg, K. E (1995). *Emotional Social Support and Cancer: A Case Study of Breast Cancer Patients' Perception of Helpful Communication Behaviors from Family and Friends.* Unpublished doctoral dissertation, University of Florida, Gainesville, FL.

Kaplan, H. S. (1992). A Neglected Issue: The Sexual Side Effects of Current Treatments for Breast Cancer. *Journal of Sex and Marital Therapy, 18*, 3–19.

Kasper, A. S (1994). A Feminist, Qualitative Methodology: a Study of Women with Breast Cancer. *Qualitative Sociology, 17*, 263–281.

Katz, P. R (1993). Health Education in a Cultural Context. *Journal of the American Academy of Physician Assistants, 6*, 478–481.

Lauver, D. (1994). Care-seeking Behavior with Breast Cancer Symptoms in Caucasian and African-American Women. *Research in Nursing and Health, 17*, 421–431.

Lauver, D., Coyle, M., & Panchmatia, B. (1995). Women's Reasons for and Barriers to Seeking Care for Breast Cancer Symptoms. *Women's Health Issues, 5*, 27–35.

McGuire, W. L., Tandon, A. K., Allred, D. C., Chamness, G. C., Ravdin, P. M., & Clark, G. M. (1992). Prognosis and Treatment Decisions in Patients with Breast Cancer Without Axillary Node Involvement. *Cancer, 70*, 1775–1781.

Mills, M. K., & Blank, R. H. (Eds) (1992). *Health Insurance and Public Policy: Risk, Allocation, and Equity.* Westport, CT: Greenwood Press.

Neale, A. V., Tilley, B. C., & Vernon, S. W. (1986). Marital Status, Delay in Seeking Treatment and Survival from Breast Cancer. *Social Science and Medicine, 23*, 305–312.

Neuling, S. J., & Winefield, H. R. (1988). Social Support and Recovery After Surgery for Breast Cancer: Frequency and Correlates of Supportive Behaviors by Family, Friends and Surgeon. *Social Science and Medicine, 27*, 385–392.

Pelton, R., & Overholser, L. (1994). *Alternatives in Cancer Therapy: The Complete Guide to Non-Traditional Treatments.* New York; Simon and Schuster, Inc.

Peters-Golden, H. (1982). Breast Cancer: Varied Perceptions of Social Support in the Illness Experience. *Social Science and Medicine, 16*, 483–491.

Polissar, L., & Finley, M. L. (1985). Time Trends and Key Factors in the Choice of One-Step or Two-Step Biopsy and Surgery for Breast Cancer. *Social Science and Medicine, 21*, 733–740.

Reaby, L. L., Hort, L. K., & Vandervord, J. (1994). Body Image, Self-concept, and Self-esteem in Women Who Had a Mastectomy and Either Wore an External Breast Prosthesis or Had Breast Reconstruction and Women Who Had Not Experienced Mastectomy. *Health Care for Women International, 15*, 367–375.

Roberts, C. S., Cox, C. E., Shannon, V. J., & Wells, N. L. (1995). A Closer Look at Social Support as a Moderator of Stress in Breast Cancer. *Health and Social Work, 19*, 157–164.

Satariano, W. A., & DeLorenze, G. N. (1996). The Likelihood of Returning to Work after Breast Cancer. *Public Health Reports, 111*, 236–242.

Schoen, R. E., Marcus, M., & Braham, R. L. (1994). Factors Associated with the Use of Screening Mammography in a Primary Care Setting. *Journal of Community Health, 19*, 239–253.

Stellman, S. D. (1987). *Women and Cancer.* New York: Harrington Park Press.

Strax, P. (1974). *Early Detection: Breast Cancer is Curable.* New York: Harper and Row Publishers.

Street, R. L., & Voigt, B. (1997). Patient Participation in Deciding Breast Cancer Treatment and Subsequent Quality of Life. *Medical Decision Making, 17*, 298–306.

Suarez, L. (1994). Pap Smear and Mammogram Screening in Mexican-American Women: The Effects of Acculturation. *American Journal of Public Health, 84*, 742–746.

Venet, L. (Ed.) (1979). *Breast Cancer: A Practical Guide to Diagnosis and Treatment.* New York: SP Medical and Scientific Books.

Virag, I. (1995). *We're All in This Together: Families Facing Breast Cancer.* Kansas City: Andrews and McMeel.

Waxler-Morrison, N., Hislop, T. G., Mears, B., & Kan, L. (1991. Effects of Social Relationships on Survival for Women with Breast Cancer: A Prospective Study. *Social Science and Medicine, 33,* 177–183.

Webber, P. A., Fox, P., Zhang, X., & Pond, M. (1996). An Examination of Differential Follow-up Rates in Breast Cancer Screening. *Journal of Community Health, 21,* 123–133.

Weinberg, N., Uken, J., Schmale, J., & Adamek, M. (1995). Therapeutic Factors: Their Presence in a Computer-mediated Support Group. *Social Work With Groups, 18,* 57–69.

Weiss, L. (1997). *Private Medicine and Public Health: Profit, Politics, and Prejudice in the American Health Care Enterprise.* Boulder, CO: Westview Press.

Wilkinson, S., & Kitzinger, C. (Eds) (1994). *Women and Health: Feminist Perspectives.* London: Taylor and Francis Ltd.

Zapka, J. G., Harris, D. R., Stoddard, A. M., & Costanza, M. E. (1991). Validity and Reliability of Psychosocial Factors Related to Breast Cancer Screening. *Evaluation and the Health Professions, 14,* 356–367.

THE SOCIAL CONSTRUCTION OF SEX CATEGORIES AS PROBLEMATIC TO BIOMEDICAL RESEARCH: CANCER AS A CASE IN POINT

Barbara Hanson

ABSTRACT

This chapter uses a feminist social construction argument to question the use of sex categories in biomedical research. It is argued that dichotomous sex categories reflect social pressures to categorize and create difference, gender, as much or more than physical sex characteristics in bodies. Because a dichotomous categorization scheme is overlayed on a continuous physical phenomenon contradictions appear when sexual artifacts like hormones, genitalia, sexuality, sexual activity, and reproduction are attributed with causal status either in causing/curing, increasing/decreasing risk. The implications of this argument are illustrated using a content analysis of medical writing about cancer since 1900. Implications for definition of women's health and pan-sex biomedical research are discussed.

INTRODUCTION

Are social assumptions about physical sexual differences legitimate as categories for biomedical research? I argue that they are not on two

Health, Illness, and Use of Care: The Impact of Social Factors, Volume 18, pages 53–68.

ISBN: 0-7623-0740-4

epistemological grounds: the imposition of a dichotomy on a continuum, and the use of an assumption as a conclusion. These aspects of constructing a biology of sex difference based on social assumptions lead to modes of interpretation of evidence directed at the centrality, objectification, and pathologization of female aspects of sex.

This argument is grounded theoretically in constructivism and feminist thinking. From constructivism I take up the notion that people act based not on things but rather the meaning things have for them. Whereby physical biology is fixed, the interpretations of this physical entity are open to social construction which may be related in varying degrees to any physical object. From feminist thinking that has held the specific content of social constructions of biology up to critical scrutiny (Fausto-Sterling, 1985; Bleier, 1984; Keller, 1985), I take up the idea that there are social constructions of difference that are attributed to alleged biological differences. These analyses have targeted the way social discrimination has been justified by biomedical models of difference. However, they have yet to question the construction of the biological categories internally to biomedicine, often taking these as a given.

By looking at how biology is constructed in biology, it is possible to argue that criticism of biological reductionism at the level of social constructions which assumes biological sex differences as a given (Yanagisako & Collier, 1990; Dupre, 1990) is too narrow. Or, social analysis has been too hesitant to enter into critique of categorization within biology and medicine (Bordieu, 1989). This ties my analysis to postmodernist conceptions that race and gender are "too reductive to describe the complexity of social identities" (Gagnier, 1990: p. 23). Biomedical constructions are as prone to mediation and creativity as an set of social meanings. Therefore, they require critical examination.

It may be that the social constructions of biomedical sex classifications have been too few. A sexual dichotomy is an epistemological stance that plays out in ways of collecting, interpreting, and acting upon on research data. Taking cancer as a case in point, it is possible to argue that sexual difference more often focuses on female difference, and the assumptions about such differences spawn courses of action that may be detrimental to women.

This theoretical argument is grounded in a content analysis of medical writing about cancer since the early 1900s. My curiosity was piqued by two observations. Long term survival of cancer has not increased since the 1950s (Beardsley, 1994; Bailar & Smith, 1986; American Cancer Society, 1967). The death rate from breast cancer has remained stable since the 1940s (Tormey & Cohen, 1979; Petty & Park, 1979) or 1930s (Garfinkel, 1991). I argue that one reason cancer research and intervention appears stalled is widespread use of a problematic sex classification as a major control/comparative variable. Since the control category of dichotomous sex is problematic so too are conclusions and intervention that are based on it.

I began my research with a broad overview of sources on cancer/oncology, taking in books on a variety of areas of cancer study – hormones, surgery, radiology, and various types of cancer focusing first on sex This initial broad overview involved several hundred books and led to further purposive sampling of sources focusing on: (a) summaries of risk factors among all cancer sites, and (b) intervention strategies and risk factors for cancer sites where focus seemed relevant.

In total over 2,500 sources were reviewed. Of this total over 1600 of were from medline, the computerized version of Indexus Medicus, focusing on 1990 to the present. All the medline sources were reviewd at the abstract level and key articles were reviewed in full. Of the 2,500 total sources over 300 were entered into a database for citing. This subset includes texbooks on cancer and oncology, articles on medical research findings, abstracts from "Medline," popular media sources, books and articles from the sociology of health and a broad spectrum of applied disciplines discovered through "Sociofile" which goes back to 1974.

I have chosen to be selective along the particular theme of sex categorization and how it comes into the definition, collection, and interpretation of data about cancer. Precedents for this type of analysis are found in work on sex determination (Findlay, 1995), reproductive technology (Rowland, 1992), and menopause (McCrea, 1983).

EPISTEMOLOGICAL QUESTIONS ABOUT USING SEX CATEGORIES

How do we know what we think we know about sex categories? I offer that there are widespread social beliefs about sexual differentiation that mediate biomedical interpretations of evidence. Gender classification is one of the most fundamental and powerful categories in social behaviour. So pervasive are these beliefs that they are taken for granted and rarely given any explicit attention in biomedical research. There are two aspects to the taken for granted social construction of sex categories that bear consideration – its dichotomous nature and its role as both assumption and conclusion.

Dichotomy

It is possible to argue that while physical sexual characteristics exist on a continuum more and less and bigger and smaller, sex categories are either/or, yes/no. Any given physical body is neither 100% female nor 100% male. We all have relative sexual characteristics, are more or less male or female. More or less body hair. Bigger or smaller genitalia. Levels of both estrogen and testosterone, vary not just from person to person but from hour to hour in any given body. Sex characteristic development is a complex interaction of genetic, hormonal and environmental factors.

Gender, the social distinction, is however, a dichotomy with two categories. Even though there are many gendered sexual identities, there is still a basic male/female, man/woman classification. People are usually put into one category or the other based on the presence or absence of penis at birth or more currently pre-natal ultrasound. This social separation into a presumed dichotomy does not however reflect the continuous nature of biological attributes. Further, so strong is the power of this social categorization that it can override the physical biological attributes in the case of hermaphrodites – people born with ambiguous genitalia, socialized to one category or the other who developed a more pronounced biological sex at puberty. In cases where socialization has been the opposite of the biological, it is usually easier to alter the biological than the social.

This leads to an argument that it is social construction as much or even more than physical entity that leads to sorting bodies into either male or female. This begins with the first idea of having a child, through ultrasound, amniocentesis, to birth. Sex classification a social decision based on the presence or absence of a penis at birth, or more recently, fetal ultrasound. In this sense a gendered construction of sex is a yes/no categorization. We define boys and non-boys, rather than boys and girls. Where sex is not clear there is pressure to decide (Findlay, 1995). Here it is important to remember that dichotomous sex is a relatively recent idea, taking hold in industrializing Europe as an alternative to the classical conception of relative sex. Bodies were historically considered relatively male (Laqueur, 1987). Ovaries were testicles turned inward. Babies came whole in sperm. The vagina was an inverted penis.

In Europe in the 1800s a model of difference based on kind rather than degree emerged and remains to this day. Instead of males and females having the same characteristics to greater and lesser degrees they were thought to be two different kinds of entities (Laqueur, 1990). Although it can be argued that this was based on increasing scientific knowledge, it can also be argued that this social construction of separation in biomedicine reflected growing pressure toward social difference that would justify a growing division of labour and sexual double standard.

I agree with the latter interpretation and argue that it demonstrates the ability of social constructions to influence biomedical constructions of physical entities. Sexual dichotomy remains with us in social constructions despite growing sophisticated biomedical knowledge. Sexual differentiation, while a social belief, is difficult to justify in physical terms. Chromosomes the so called "gold standard" for determining sex are becoming less relevant (Edlin, 1990) as it becomes possible to manipulate hormone levels with drugs in order to promote development of more pronounced sex attributes or even the opposite sex attributes. Genetic males can have babies. Post menopausal women can have babies. Sex is thus a theoretical ideal of male and female that specific bodies approximate but no one body is purely one or the other.

However, medical research by and large takes these socially constructed, gendered, sex categories unproblematically to represent a biological difference. This has consequences for comparison. Sex becomes a control or comparative category for research. Men and women are assumed to represent the theoretical ideal of male and female. Conclusions are then drawn about how sex attributes like estrogen or percentage body fat enter into a condition like cancer. This is a false comparison since the biological attributes of the people in the two categories are continuous and never tested. Rather, they are assumed to represent a dichotomous difference.

This makes it possible to argue that a dichotomous sex category is not a direct reflection of the physical makeup of human bodies. It echos a social belief in separation more than a detectable physical separation. Physical sexual characteristics exist on a continuum, sex categories are a dichotomy. The imposition of a dichotomy on a continuum is therefore like dividing mercury with a ruler.

It becomes problematic for biomedical research and intervention because the phenomena reconstitutes itself into a whole when the ruler is removed, where the separation cannot be enforced. This may be why sex categories in research show a pattern whereby artifacts of sexual differentiation can both cause and cure the same thing. It is a separation forced on something that is not legitimately separable. So it is not surprising that evidence is either contradictory or one study proves, the next study disproves. Biomedical research is using an inappropriate separation that may have more social than physical relevance, and is therefore doomed to be trapped in an inescapable circle of contradiction. Further, where sexual difference is used to define a condition it is prone to pathologizing artifacts of sexual difference. Since sexual difference or object is more often female centric, sexual dichotomy may be more detrimental to women.

Tautology

While physical sexual characteristics are given, their sex classification is not. Therefore imposing sexual dichotomy then attributing cause to sexual difference is using sex categories as both assumption and conclusion. This is the definition of tautology. A difference is assumed then this difference is concluded to be the source of observed differences.

As argued above, the assignment of sex categories is something that is neither clear nor fixed. It is up for grabs. Witness the ways sex categories show historical and cultural variation. Sex classification is subject to human group interpretation. The content of such interpretations therefore needs to be held up to critical attention. If dichotomy serves a social belief as much or more that a physical presence it must be allowed that this a human group decision, whether implicit or explicit. It is taken as true, rather than tested, therefore an assumption. The shape and content of this "taking as" process it was must be examined.

Once we see that sex categories are something tied to taken for granted social constructions of difference rather than some concrete physical difference, it must be offered that sex categories are an assumption. As such they form a tautology when attributed with causal significance. The logic becomes sex, therefore sex.

Put another way sex is acting as both dependent and independent variable. However, since it is being held constant (dependent) it is not free to vary (independent) in practice. Therefore, when the sex category or variable is used in research and intervention it is prone to spawning various form of contradictory information and actions which reinforce social constructions of beliefs about sex difference.

Like light through a prism, the relational complexity of human group experiences is refracted into social categories, one set of which is gender. Health research as co-emergent in this social context imposes social categories on biological phenomena without ever questioning whether or not they are legitimate biologically. Construction of sex significance comes only after the fact when social presumptions are used to explain differences between the bodies in the categories. Talk of hormones and genitalia only comes after the social gender categories of man and woman are used to sort a sample and look for differences attributed to maleness versus femaleness. Gender categories are assumed to differentiate biologically but biological evidence does not enter into the initial sorting process.

Social presumptions about the relationship between sex and gender govern the medical decision to sort. However, the construction of illness consequences and risk factors harks to the biological rather than social differences between the people in the categories. While biological significance is the assumption, social construction is the practice. This is a process of biological reductionism in that the relational character of human group experience is disregarded in favour of presumptions about the individual nature of biology as bounded by skin. Research talks about sex differences but uses social distinctions to draw comparisons.

Blindness to a gender argument, sex categorization is mediated by social construction of gender, means that health research and practice is effectively sexist. It constructs a view of biology based on social prejudices rather than physical evidence. Therefore, any conclusions about health and illness which begin with the imposition of a gender dichotomy on a biological continuum are bound to distort. In the course of definition, detection, and intervention, in cancer, gender, the social construction of experiences, is being explained based on sex, biological differences surrounding hormones and reproductive organs. But this relationship can't be legitimately considered direct when it is examined from the theoretical position of the social construction of gender. It is a process of biological reductionism. Gender is reduced to sex, then explanation and intervention is based on sex. This is tautological in that the argument is sex therefore sex. Research and intervention founded in such a tautology is destined to contradiction, impasse, and repetition.

MODES OF INTERPRETATION SPAWNED BY THE ASSUMPTIONS UNDERLYING SEX CATEGORIES

If we see the world in terms of sexes we will explain the world in terms of sex. Since threads of sexism run through the social context in which health research emerges, sex focus leads to female focus. Male sex tends to be unnoted, underplayed, or when noted, attributed with positive influence. Femaleness is forefront. Reducing women to their femaleness, is a way of constructing them as objects in male dominated biology and medicine (Rowland, 1992) and distancing them from their bodies (Martin, 1987). This means women have trouble defining themselves as persons, not just in terms of their now objectified body, but in terms of the few body parts which now form their identity.

I use a conception of sexism as sex focus, a world view or epistemology which sees the world in terms of sex distinctions. This stance is echoed in Chodorow's work which takes gender and sexuality as problematic (Chodorow, 1990). Within this it is possible to examine critically not just discrimination against women, but how sex focus leads to particular sets of categories which have an impact on how all people are sorted. The importance of this distinction in medical practice is illustrated in Mitchinson's historical analysis of women and their doctors in Victorian times, noting the attribution of sex cause to deaths easily explained by gender differences (Mitchinson, 1991).

It may seen less offensive when a health practitioner reduces a woman to her body then to breast, cervix, uterus, or hormones, than when a prospective employer or Hollywood film does. However, the social construction of women in terms of artifacts of female sex in medical research and practice may be just as dangerous for women because it reiterates the biological reductionism inherent in justification of social differences. One of the key problems is the way in which the body is reduced not just to its biological/medical identity, but to a few of its parts – sex things. Reducing the person as a whole, first to body, then to some body parts makes the process of dealing with health and illness prone to sins of inattention.

Female Sex as Pathological

One of the most important thrusts of feminist health has been to argue that there has been a general tendency to pathologize womaness in health care (McCrea, 1983; Clarke, 1990; McKinlay, 1972). This point is illustrated in a range of issues like drug use (Harding, 1986), reproductive technology (Ford, 1986), (Rowland, 1992), physical exercise (Vertinsky, 1990), menopause (McCrea, 1983) and my own work on the myth of biological time clock (Hanson, 1997). In terms of care providers, Ehrenreich and English chronicle the process of marginalization, even criminalization, of women's health practices (Ehrenreich & English, 1973). I list here only a few highlights of a vast and growing body of work which suggest that sex focus in health pathologizes women.

When biomedical focus becomes sex centric concrete problems for women emerge. The tendency to focus on women's sex related body parts means that women are sexual objects. This is as troublesome in medicine as in media. To centre on female body parts and either glorify them as the zenith of sexual allure or degrade them as cancer cites is equally dangerous – two sides of the same coin. Focus on female sex differentiators leads inevitably to the objectification and commodification of female body parts. This is as true of cancer as reproductive health (Rowland, 1992).

Female sex emphasis is reflected in the promotion and practices of detection focused on genitalia, sexual development, and sexual activity. Both sexes are routinely examined from puberty on for possible tumours in accessible genitalia, testicles, prostate, penis, breasts, vagina, and cervix. Tumours of the liver, pancreas, lymph glands, lung, kidney, or other less accessible organs or systems are not. This opens up the possibility that genitalia focus in cancer may be as much a case of what is routinely screened rather than what is ultimately most dangerous.

Mammography, the compression of breasts with paddles in order to x-ray them and look for possible malignant growths has several troubling dimensions. First, the process is painful. Pressure applied to breasts in order to take the x-ray is a vague range (Eklund, 1991; Sullivan, Beam, Goodman & Watt, 1991), and in practice is usually the amount of pressure women can stand. The greater the better. Second, where a physician suspects cancer, even a negative (cancer not suspected) mammogram, is insufficient reason to not proceed to biopsy (Hendee, Manaster, Harnsberger, Bragg, Thompson & McClennan, 1991). Third, recent research and review of research discussion suggests that there may be no benefit to mammography, or it may only be beneficial after age 50 (Dotta & Klein, 1993; Scanlon, 1991) and perhaps not over 64 (Davidson & Sacks, 1995). Finally, breast lump detection may have no relevance to outcomes in cancer (Tormey & Cohen, 1979; Khafif, Rafla & Kopel, 1986). Breast cancer mortality rates have remained relatively stable since the 1930s despite rapid increases in the percentage of women who have mammograms (Garfinkel, 1991).

This makes mammography troublesome for several reasons. It reinforces a breast first view of women's health. It shifts the responsibility for breast cancer onto women. One author suggests that women have been blamed for breast cancer (Read, 1995). I concur and add that shifting culpability from a women to some of her body parts, those associated with female sex, continues this process. The questionable effectiveness of mammography for detecting cancer, except in a limited age range, and tendency for false positive readings leads to high levels of anxiety and large numbers of breast biopsies, than a false sense of security with negative results. This focuses attention on women as breasts while justifying the costs both in terms of physical and emotional suffering, and health care costs. It shifts responsibility for cancer to women by making them responsible for initiating and complying with the detection process. Futher, it may give a sense of progress because short-term survival has increased. However, this may be an artifact of earlier detection since mortality rates have been stable since the 1930s or 1940s.

PAP smears are based on the theory that on the way to becoming cancerous, cervical cells move through stages of abnormality which can be detected. If detected in early stages the abnormal area can be removed by surgery or laser treatment. However, no clinical trials were done to confirm this (Kink, 1991). Further, the PAP test fails to meet the criteria for mass screening programs (Foltz & Kelsey, 1978).

Breast self examination (BSE) never been subjected to a randomized clinical trial to test its effectiveness either (Mant, Vessey, Neil, McPherson & Jones, 1987). Even though it has no legitimacy as a screening mechanism breast self examination it is publicized as health promotion (Moss, 1995). Women are drawn into complicity with the pathologization of femaleness and held responsible through emphasis on early detection.

The contradictions and paradoxes of sex centric explanations cancer are further illustrated by looking at male cancer, which is coded by a female centric focus. Estrogen is seen as good for the treatment of prostate cancer (Karr & Murphy, 1985). It is seen as a standard treatment for prostate cancer in one study (Morse & Whitmore, 1985), while another in the same edited text points out that emphasis has been placed on the prostate and hormones when distant metastases kill the patient (Karr & Murphy, 1985). This raises the question of why we call it prostate cancer and treat prostate cancer. I would argue that so pervasive is sex focus that it colours these decisions. Even though it is cancer in some other organ which is fatal, the focus is sex based. If we see the world in terms of sex, we will explain the world in terms of sexes?

The Janus Like Face of Estrogen

One of the directly observable offshoots of my argument is the Janus like face of hormones. Female hormone, estrogen, is cast in the light of both hero and villain, depending on the stage of the female life cycle. Even this is contradicted, but the overriding principle is that female hormones are bad after puberty but become good at menopause. So femaleness is bad any time after puberty. Women's sex hormones cause problems from puberty to menopause but then cure problems. By definition then, there is something inherently wrong, unhealthy, carcinogenic, about femaleness for all phases of the life course after puberty.

This argument defies internal resolution in that it requires a physical biology which is transmutable, based on different principles at the same time. A universal extension in terms of gravity would be that when you drop something on earth, sometimes it falls down, sometimes up. Similarly "magical" hormones, targeted by the lens of individual sex units can both promote and cure cancer. In medical writing estrogen has repeatedly been attributed with the ability to transmute itself as alternatively cause and cure of cancer.

A review article evaluating evidence on hormones in general, relative to cancer, concludes that the best evidence among "all human tumors" is for the DES and genital tract cancer (Meakin, 1979). Recent findings indicate that combined estrogen/progesterone therapy raises breast cancer risks (Colditz, Hankinson & Junter et al., 1995).

Estrogen seems to be alternatively bad/good for breast cancer depending on whether or not there is estrogen in the system. Estrogen is bad, or causal of cancer, evidenced by the use of "anti-estrogen" or androgens to increase disease free survival of breast cancer in pre-menopausal women (Pearson, Gordon, Hubay & McGuire, 1985: p. 493). More recently tamoxifen, an "anti-estrogen" has been used to treat pre-cancerous breast lumps (Gibson & Jordan, 1993). Another source suggests if pre-menopausal, estrogen is bad, and recommends treatment with androgens (Tormey & Cohen, 1979: p. 437). However, it is good for treating breast cancer more than five years after menopause, but bad in the first five years after menopause (Tormey & Cohen, 1979: p. 438; Anderson & Rossof, 1979).

The flip flop pattern of good/bad estrogen shows up how the assumption of sex differences has led to explanation in terms of sex differences. Reliance on the basic social separation of gender to indicate sex and provide a fundamental set of control groups is ultimately tautological in that sex becomes functionally independent and dependent variable. The pervasiveness of this paradoxical line of explanation is evidenced in the statement "There is no question that breast cancer is a disease related to ovarian activity. . . ." (Miller, 1982: p. 231; Drill, 1979: p. 138). Yes, females do get breast cancer and have ovaries. Femaleness is therefore carcinogenic. I argue that this is tautological owing to the assumption of femaleness as an exclusive or primarily women's trait. It is also tied to a site focus approach that may mask common features of cancer for men and women.

DISCUSSION

My arguments suggest that the rationale behind the assignment of sex categories needs to be examined internally to biomedical research and intervention. Once we take sex classification as problematic rather than given then ideas that are based on this classification must be taken as problematic as well. This begins with reconsidering the definition of women's health and specifics in terms of cancer research, detection, and intervention.

The Definition of Women's Health

A gendered sexual dichotomy reflected in women's health may be dangerous for women. Defining women's health issues in terms of reproductive health or sex differences from males (Lundy & Mason, 1994; Millner & Wilderman, 1994), reduces women to a small number of their parts and bodily functions. This reinforces the biological reductionism which is at the heart of sexist ideology. It may be less obvious in medicine and biology but is none the less sexist for it. Women may have historically suffered from lack of attention to their cardiac problems. Now that women's cardiac problems are being acknowledged hormone replacement therapy is suggested as treatment. Why the tendency to sexualize women first by exclusion then by intervening sexually?

Emphasizing dichotomous sex differences and sex consequences as the raison d'etre for women's health masks the totality of women's bodies and social experiences. Women, already marginalized by socially imposed categories which are legitimized by alleged biological differences, are not likely to be well served by the reification of those differences in health care. As long as women are cast as the other or second sex (de Beauvoir, 1989) women's health will continue to be managed by men (Grossman & Bart, 1979) to the detriment of women.

The lesson here is pre-menstrual syndrome. Women demanded medical attention for this female experience. They got it. Included in the *Diagnostic and Statistical Manual of Mental Disorders*, of the American Psychiatric Association, Fourth Edition, is "Premenstrual Dysphoric Disorder" (American Psychiatric Association, 1994). Given that the majority of women experience pre-menstrual syndrome, femaleness has been defined as a Mental Disorder.

I recognize that there may have been sound political reasons for defining women's health as reproductive health, but argue that this political need to define a mandate of difference to get funds might be applied selectively in the internal workings of biomedicine and health care for women. Focus on women's health as sex related may be dangerous because it repeats the biological reductionism that goes into defining sex categories in the first place. Concentration on femaleness may be as discriminatory as focus on maleness to the exclusion of femaleness, because in both cases assumptions of a sexual dichotomy are being imposed. It begins a process toward medicalization hence pathologization of female sexual characteristics.

The founding of women's health was a necessary first step. Establishing research and practices for women in the face of the dominance of men in biomedicine brought the other side of sex in health into focus. It is now worthwhile to consider the issues of sexual biology in a full range, not just at two idealized poles.

Cancer

The tendency to gender stream detection and intervention into sex categories may have an impact all down the line to mask something which, if treated as a single entity, would show up patterns of differentiation on some other characteristic. The current practice would be to separate breast cancer and prostrate cancer patients, study and treat them differently. The paradoxes of this approach and the emphasis on female sex are suggested by research indicating that mortality rates from cancer for all cites combined is consistently greater for males (Garfinkel, 1991; Nicholson & Davis, 1990) and estimated to be seeking parity (Garfinkel, 1991). If sexual characteristics are to be held up for examination it should be artifacts of male rather than female sex. Sex focus may be distracting attention from crucial causal/risk factors found among both groups of patients such as lifestyle and environmental factors and in so doing dividing up the phenomena in ways which render it invisible.

My suspicion is that the issue of sex site focus is one part of a broader problem with site focus as an epistemological rubric for looking at cancer. Sex focus highlights the tendency to look at where cancer is found to draw conclusions. If cancer is a full body system issue, a relational issue, or an environmental issue, it is hiding in plain sight of current viewing lenses. What I suggest is that emphasis on women's health most notably in terms of activism which has led to increasing funding on research of breast cancer is a positive move in total, providing that women's health is not confused with breast health. It's time to turn breast focus into person focus. Take into account the totality of women's bodies not just how they are theorized to differ in kind from men's. It also means looking at aspects of women's social status and construction which may shape how they come to be seen by medical viewing lenses. Its time for a sociological view internal to the workings of biomedicine.

ACKNOWLEDGMENTS

Thanks to Lisa Strohschein for gathering and summarizing material. The project from which this chapter emerges was supported through York University Specific Grants, Atkinson College Grants in Aid, and my visiting appointments at the Institute for Research on Women, Rutgers University, and Department of Sociology, Princeton University. Contact the author at hansonbg@yorku.ca

REFERENCES

American Cancer Society. (1967). *New Directions in Cancer Research*. New York: American Cancer Society Inc.

American Psychiatric Association (1994). *DSM-IV Diagnostic and Statistical Manual of Mental Disorders*. Washington, D.C.: American Psychiatric Association.

Anderson, K. M., & Rossof, A. H. (1979). The Influence of Androgens in Tumor Development. In: J. A. Kellen & R. Hilf (Eds), *Influences of Hormones in Tumor Development* (pp. 1–30). Boca Raton (Florida): CRC Press Inc.

Bailar, J. C., & Smith, E. (1986). Progress Against Cancer? *The New England Journal of Medicine, 314*, 1226–1232.
Beardsley, T. (1994). A War Not Won. *Scientific American, 270*, 130–138.
Bleier, R. (1984). *Science and Gender: A Critique of Biology and its Theories on Women*. Toronto: Pergamon Press.
Bordieu, P. (1989). Social Space and Symbolic Power. *Sociological Theory, 7*, 14–25.
Chodorow, N. J. (1990). What is the Relation Between Psychoanalytic Feminism and the Psychoanalytic Psychology of Women? In: D. L. Rhode (Ed.), *Theoretical Perspectives on Sexual Difference* (pp. 114–130). New Haven: Yale University Press.
Clarke, J. N. (1990). *Health, Illness, and Medicine in Canada*. Toronto: McClelland & Stewart Inc.
Colditz, G. A., Hankinson, S. E., Junter, D. J., Willet, W. C., Manson, J. E., Stampfer, M. J., Hennekens, C., Rosner, B., & Speizer, F. E. (1995). The Use of Estrogens and Progestins and the Risk of Breast Cancer in Postmenopausal Women. *New England Journal of Medicine, 332*, 1589–1593.
Davidson, T., & Sacks, N. P. M. (1995). Principles of Surgical Oncology. In: A. Horwich (Ed.), *Oncology: A Multidisciplinary Textbook* (pp. 101–115). New York: Chapman and Hall.
de Beauvoir, S. (1989). *The Second Sex*. New York: Vintage Books.
Dotta, L., & Klein, E. (1993). Questioning Mammography. *University of Toronto Magazine*, 20–23.
Drill, V. A. (1979). Oral Contraceptives and Cancer. In: J. A. Kellen & R. Hilf (Eds), *Influences of Hormones in Tumor Development* (pp. 137–154). Boca Raton (Florida): CRC Press Inc.
Dupre, J. (1990). Global versus Local Perspectives on Sexual Difference. In: D. L. Rhode (Ed.), *Theoretical Perspectives on Sexual Difference* (pp. 47–62). New Haven: Yale University Press.
Edlin, G. (1990). *Human Genetics: A Modern Synthesis*. Boston: Jones and Bartlett Publishers.
Ehrenreich, B., & English, D. (1973). *Witches, Midwives, and Nurses: A History of Women Healers*. New York: The Feminist Press.
Eklund, G. W. (1991). Mammographic Compression:Science or Art? *Radiology, 181*, 339–341.
Fausto-Sterling, A. (1985). *Myths of Gender: Biological Theories about Women and Men*. New York: Basic Books Inc.
Findlay, D. (1995). Medical Science,Feminism and Intersexuality. *Canadian Review of Sociology and Anthropology, 32*, 25–52.
Foltz, A. M., & Kelsey, J. L. (1978). The Annual Pap Test: A Dubious Policy Success. *Milbank Memorial Fund Quarterly/Health and Society, 56*, 426–462.
Ford, A. R. (1986). Hormones: Getting Out of Hand. In: K. McDonnell (Ed.), *Adverse Effects: Women and the Pharmaceutical Industry* (pp. 27–49). Toronto: Women's Educational Press.
Gagnier, R. (1990). Feminist Postmodernism: The End of Feminism or the Ends of Theory. In: D. L. Rhode (Ed.), *Theoretical Perspectives on Sexual Difference* (pp. 21–30). New Haven: Yale University Press.
Garfinkel, L. (1991). Cancer Statistics and Trends. In: A. I. Holleb, D. J. Fink & G. P. Murphy (Eds), *American Cancer Society Textbook of Clinical Oncology* (pp. 1–6). Atlanta: The American Cancer Society.
Gibson, D. F. C., & Jordan, C. V. (1993). The Roles of Hormones (Estrogen) in Prevention and Treatment of Early Breast Cancer. In: I. M. Ariel & A. C. Cahan (Eds), *Treatment of Pre-*

Cancerous Lesions and Early Breast Cancer (pp. 112–122). Philadelphia: Williams & Wilkins.

Grossman, M., & Bart, P. B. (1979). Taking The Men Out of Menopause. In: R. Hubbard, M. S. Henifin, & B. Fried (Eds), *Women Look at Biology Looking at Women* (pp. 163–185). Cambridge: Schenkman Publishing.

Hanson, B. (1997). *Social Assumptions, Medical Categories (Supplement 1: Advances in Medical Sociology.* Greenwich, Connecticut: JAI Press Inc.

Harding, J. (1986). Mood-Modifiers and Elderly in Canada: The Medicalization of Poverty. In: K. McDonnell (Ed.), *Adverse Effects: Women and the Pharmaceutical Industry* (pp. 51–86). Toronto: Women's Educational Press.

Hendee, W. R., Manaster, B. J., Harnsberger, H. R., Bragg, D. G., Thompson, W. M., & McClennan, B. L. (1991). Oncologic Imaging. In: A. I. Holleb, D. J. Fink & G. P. Murphy (Eds.), *American Cancer Society Textbook of Clinical Oncology* (pp. 643–677). Atlanta: The American Cancer Society.

Karr, J., & Murphy, G. P. (1985). Research Advances in Prostatic Cancer. In: V. P. Hollander (Ed.), *Hormonally Responsive Tumors* (pp. 469–485). Toronto: Academic Press Inc.

Keller, E. F. (1985). *Reflections on Gender and Science.* New Haven: Yale University Press.

Khafif, R. A., Rafla, S., & Kopel, S. (1986). *Current Guidelines for the Management of Cancer.* Toronto: Academic Press.

Kink, D. J. (1991). Cancer Detection: The Cancer-Related Checkup Guidelines. In: A. I. Holleb, D. J. Fink, & G. P. Murphy (Eds), *American Cancer Society Textbook of Clinical Oncology* (pp. 153–176). Atlanta: The American Cancer Society.

Laqueur, T. (1987). Orgasm, Generation, and the Politics of Reproductive Biology. In: C. Gallagher & T. Laqueur (Eds), *The Making of the Modern Body: Sexuality and Society in the Nineteenth Century* (pp. 1–41). Berkeley: University of California Press.

Laqueur, T. (1990). *Making Sex: Body and Gender from the Greeks to Freud.* Cambridge: Harvard University Press.

Lundy, M., & Mason, S. (1994). Women's Health Care Centers: Multiple Definitions. *Social Work in Health Care, 19,* 109–122.

Mant, D., Vessey, A., Neil, A., McPherson, K., & Jones, L. (1987). Breast Self Examination and Breast Cancer Stage at Diagnosis. *British Journal of Cancer, 55,* 207–211.

Martin, E. (1987). *The Woman in the Body.* Stratford: Open University Press.

McCrea, F. B. (1983). The Politics of Menopause: The "Discovery" of a Deficiency Disease. *Social Problems, 31,* 111–122.

McKinlay, J. B. (1972). The Sick Role – Illness and Pregnancy. *Social Science and Medicine, 6,* 561–572.

Meakin, J. W. (1979). Hormones as Carcinogens in Humans. In: J. A. Kellen & R. Hilf (Eds), *Influences of Hormones in Tumor Development* (pp. 195–200). Boca Raton (Florida): CRC Press Inc.

Miller, A. B. (1982). Cancer of the Breast. In: K. Magnus (Ed.), *Trends in Cancer Incidence* (pp. 231–234). New York: Hemisphere.

Millner, L., & Wilderman, E. (1994). Women's Health Issues: A Review of the Current Literature in the Social Work Journals. *Social Work in Health Care, 19,* 145–172.

Mitchinson, W. (1991). *The Nature of Their Bodies: Women and Their Doctors in Victorian Canda.* Toronto: University of Toronto Press.

Morse, M. J., & Whitmore, W. F. (1985). Clinical Management of Advanced Prostatic Cancer. In: V. P. Hollander (Ed.), *Hormonally Responsive Tumors* (pp. 431–468). Toronto: Academic Press Inc.

Moss, S. (1995). General Principles of Cancer Screening. In: J. Chamberlin & S. Moss (Eds), *Evaluation of Cancer Screening* (pp. 1–13). New York: Springer.

Nicholson, W. J., & Davis, D. L. (1990). Analyses of Changes in the Ratios of Male-to-Female Cancer Mortality. A Hypothesis Generating Exercise. *Annual of the New York Academy of Science, 609*, 290–299.

Pearson, O. H., Gordon, N. H., Hubay, C. A., & McGuire, W. L. (1985). Estrogen Receptors and Prognosis. In: V. P. Hollander (Ed.), *Hormonally Responsive Tumors* (pp. 487–503). Toronto: Academic Press Inc.

Petty, W. M., & Park, R. C. (1979). Gynecologic Malignancies. In: P. J. Bergivin, J. Blom, & D. C. Tormey (Eds), *Guide to Therapeutic Oncology* (pp. 399–430). Baltimore: Williams & Wilkins.

Read, C. (1995). *Preventing Breast Cancer: The Politics of an Epidemic*. New York: Pandora an Imprint of Harper Collins Publishers.

Rowland, R. (1992). *Living Laboratories: Women and Reproductive Technologies*. Bloomington: Indianna University Press.

Scanlon, E. F. (1991). Breast Cancer. In: A. I. Holleb, D. J. Fink, & G. P. Murphy (Eds), *American Cancer Society Textbook of Clinical Oncology* (pp. 177–193). Atlanta: The American Cancer Society.

Sullivan, D. C., Beam, C. A., Goodman, S. M., & Watt, D. L. (1991). Measurement of Force Applied During Mammography. *Radiology, 181*.

The Times [Trenton] (1993). Indiana Clinton and the Temple of Doom. (1993)a23

Tormey, D. C., & Cohen, P. (1979). Breast. In: P. J. Bergivin, J. Blom, & D. C. Tormey (Eds), *Guide to Therapeutic Oncology* (pp. 431–448). Baltimore: Williams & Wilkins.

Vertinsky, P. (1990). *The Eternally Wounded Woman: Women,Doctors and Exercise in the Late Nineteenth Century*. New York: Manchester University Press.

Yanagisako, S. J., & Collier, J. F. (1990). The Mode of Reproduction in Anthropology. In: D. L. Rhode (Ed.), *Theoretical Perspectives on Sexual Difference* (pp. 131–141). New Haven: Yale University Press.

WOMEN, MEDICINE, AND SOCIOLOGY: THOUGHTS ON THE NEED FOR A CRITICAL FEMINIST PERSPECTIVE

Deborah Plechner

ABSTRACT

The relationships between women, health, and medicine are complex and contradictory. During the second-wave of the women's movement, feminists struggled to bring women's health issues to the fore. Today, their success is documented by the growing numbers of women practicing medicine, and by the increasing attention and resources devoted to women's health issues. Yet feminists remain critical of the highly gendered nature of medicine and its contribution to social inequalities. Feminists working both from within and outside the growing subfield of medical sociology have used one of its key concepts – medicalization – to explicate the negative consequences of institutional medicine for women.

The continuing medicalization of women's lives is related to key ideas about the body and important trends in the structure of medicine, particularly the growing importance and sophistication of technology. The argument is made that some instances of medicalization, including women's legal punishment for fetal abuse and coerced sterilization, herald a new medico-legal alliance that impacts the poorest of women most severely. Feminists relate the growth of the public health paradigm of

Health, Illness, and Use of Care: The Impact of Social Factors, Volume 18, pages 69–94.

ISBN: 0-7623-0740-4

medicine to the emergence of the medico-legal alliance in that both rely on the power of the state and represent the continuing medicalization of women's lives. Based on these insights, the need for a continuing critical and feminist sociological understanding of medicine is stressed and possible lines of inquiry are set forth.

> Perhaps the most significant continuing contribution of sociology to the debate about women's health care and research is the assertion that health research takes place in a context in which health, illness, gender, and science are socially constructed. Although many in the research establishment believe and would like the public or Congress to believe that science (including health research) is objective and value-neutral, a more critical medical sociology can productively argue otherwise. Like any other institutions or cultural form, the organization of science and medical science must be understood to be influenced by the particular social and historical circumstances in which it is experienced (Auerbach & Figert, 1995: 127).

The above quote by Judith Auerbach and Anne Figert speaks to the necessity of examining the relationships between women, health and medicine within a sociological framework. They assert that medical sociology and the insights it has generated about women's health as a gender issue are crucial to policymakers, medical professionals and women in general.[1] The goal of this essay is to bolster their claim by illustrating the usefulness of a critical sociological perspective for considering women's experiences with medicine as a social institution.

MEDICAL SOCIOLOGY

Medical sociology, otherwise known as the sociology of health and illness, had its beginnings almost five decades ago. At this time, the field was primarily concerned with studying the social organization of care, illness behavior and conducting empirical investigations for medical professionals. Today the sociological study of health and illness makes four central assumptions. First is the recognition that understanding health and illness requires more than biological knowledge. Second is the insight that health and illness occur within social, political, economic and cultural structures. Third, medical sociologists recognize that the institution and practices of medicine themselves must be taken as objects of inquiry and critique. And fourth, a sociological perspective on health and illness involves studying the institution of medicine and going beyond it (Charmaz & Paterniti, 1999: 1). Judith Lorber, a leading feminist sociologist, recently published a book that exemplifies a sociological *and* feminist approach to the study of illness (1997).

Lorber begins her study by pointing out that while illness and health are considered to be physiological states of being, the process of being ill is social in that it disturbs our lives. And illness is also social in the sense that across societies the "symptoms, pains and weaknesses called illness (1)" are shaped by cultural and moral values, experienced through various social interactions, and influenced by beliefs about health and illness. The process that transforms "physiological symptoms into diagnoses (1)" is heavily influenced by power differences and moral judgements. Categories that produce differences in social worth and power – gender, race and ethnicity, social class, sexual orientation and type of illness – influence research priorities, treatment modes and the production of medical knowledge.

The language of illness and medical knowledge stems from science. Doctors, nurses and health care professionals are trained in "medical science (2)." The basis of medicine as a science is the biomedical model of disease. This view assumes that disease or illness is a deviation from normal biological functioning, that diseases have specific causes that can be located in the person's body, that diseases have the same symptoms and outcome in any social situations, and that medicine is a socially neutral science (Mishler, 1981). Lorber notes many criticisms of the underlying assumptions of a medical model of science. Most important of these may be the idea that the values underlying medical science are deeply biased by the practice situations and the social characteristics of the most powerful group of medical professionals – doctors (1997: 2).

Lorber's focus on the power of medical professionals and medicine as a social institution to influence what counts as an illness, who is determined to be ill, and how the ill person is treated by others in society is at the heart of a critical medical sociology. Those working within this tradition have developed a key concept – medicalization – to describe the sociocultural process by which phenomena previously defined in non-medical terms are seen as medical problems. Peter Conrad is one of the architects of the sociological understanding of medicalization. According to him, the term first "crept" into the social scientific literature in the 1970s, both in criticisms of psychiatry and in Ivan Illich's influential critique of medicine (1976) which spoke of the overall "medicalization of life" (Conrad, 1992: 210). A number of sociological case studies – child abuse (Pfohl, 1977) and hyperactivity in children (Conrad, 1975) – revealed the usefulness of the concept and led Conrad, along with Schneider, to develop it more fully (1980).

Medicalization

The process of medicalization involves ascribing medical definitions and understandings to problems that were previously not seen as medical issues. When a social phenomena becomes medicalized it is defined in medical terms or as a distinct medical problem, treated using medical interventions or understood within a medical framework. This process does not necessarily involve medical professionals and may or may not be the result of intentional professional expansion.

Medicalization can occur on three distinct levels: the conceptual, which involves the definitional issues described above, the institutional, which involves an organization consciously adopting a medical approach to a problem, and the interactional, which may occur as part of the doctor-patient interaction. There is always the chance for resistance to the process of medicalization: "Subcultures, groups, or individuals may vary in their readiness to apply, accept, or reject medicalized definition" (Cornwell, 1984). But overall, sociologists have focused on the process of medicalization as an expansive one that often leads to medical forms of social control.

In particular, early theorists argued that medical social control would expand and replace other forms. Conrad argues that overall this has not happened but that medical social control has continued to expand. In 1979, he distinguished three types of medical social control: medical ideology, medical collaboration and technology. Medical ideology becomes a force of social control as the medical model is increasingly imposed on social life. Medical collaboration refers to the process by which doctors act as information providers, gatekeepers, institutional agents and technicians. And medical technology in the form of drugs, surgery, and genetic and other types of screening acts as a form of social control (1992: 216).

Drawing on the work of Foucault (1965, 1973), Conrad later conceptualized another form of medical social control – surveillance. This is the process by which certain conditions or behaviors come to be seen through a "medical gaze," thus enabling physicians to legitimately lay claim to all activities concerning this condition. The consequences of the process of medicalization include the domination of the problem by medical experts, the individualization and depoliticization of social problems, an overriding assumption of medical moral neutrality, the legitimation of the use of powerful medical technologies and the potential impact on public opinion and social policy (1992).

Overall, the study of medicalization has centered on the role of physicians, but they are not always involved, and can, and do, resist the process. Sociologists remain skeptical of the process but Conrad warns that we must be cautious about inferring a social control capacity. The process of medicalization can also work the other way: that is, medical conditions previously seen as social problems – for example, homosexuality or physical disability – can be demedicalized as the result of either an organized movement or technological advances. And there has been very little cross-cultural research or any research for that matter that goes beyond isolated case studies (Conrad, 1992).

Despite these shortcomings the concept of medicalization has been crucial in the development of a critical medical sociology. Sociologists have used the concept to analyze both *deviant behaviors* – madness, alcoholism, opiate addiction, hyperactivity and learning disabilities in children, eating disorders, child abuse, compulsive gambling, and infertility – and *natural life processes*, including sexuality, childbirth, child development, menstrual discomfort, menopause, aging and death (Conrad, 1992: 213). Critical medical sociologists were quick to point out race and class biases in Western medicine (Lorber, 1997). But medical sociologists have ignored issues of gender and women's health (Auerbach & Figert, 1995). It is feminists who have utilized the concept to reveal the overwhelming degree of medicalization in women's lives, and how the practices and imagery of medical science contribute to the social construction of gender.

The feminist critique of medicine has taken place on many levels. Briefly I would like to step back in time and examine the history of the movement which helped to construct *women's health* as a social and political issue. This history is important to understanding where we are today in terms of women's health and in terms of our sociological understanding of medicine, health and illness as social constructions. Feminist insights have helped to delineate and shape several key issues including the degree of medicalization in women's lives.

THE WOMEN'S HEALTH MOVEMENT(S)

During the 1960s and 1970s feminists began to examine the ways in which medicine produces diagnoses and treatments that are harmful to women, depicts women in textbooks and scholarly reports in stereotypical and negative ways, and is not objective and value free (Bell, 1995).[2] Disparate and sexist health care was a major concern of second wave feminists in North America, Australia, New Zealand and Europe. These feminists used grassroots organizing to establish many women's health centers. Today many of these centers, along with grass-roots feminism, have declined but women's health remains a key issue for Western feminists. Judith Auerbach and Anne Figert (1995) along with Leslie Doyal (1996) note that it is incorrect to speak of any one women's health movement. Within the U.S. and throughout the world, there are many movements working towards the goals of reproductive self-determination, affordable, effective and human health care, the satisfaction of basic needs, a safe workplace and physical security. While there are growing links between national and international, western and third-world movements, in general the priorities and strategies of these movements remain necessarily different.

In general, women's access to health care varies worldwide based on individual wealth or poverty and the cost and availability of services in the local community. The situation of third-world women is basically one of material deprivation while women in industrialized nations suffer from over-exposure to medicine and medical technology (Doyal, 1996). In less-developed countries women still struggle for basic reproductive autonomy and suffer grave discrepancies in their nutrition, health and education when compared with men (Joel, 1997). Even within the U.S. there are huge disparities in access to health care among women. For instance, the rapid spread of AIDS among Black and Hispanic women is related to their increased chances of poverty and the fact that health centers for poor women are rapidly being dismantled across our nation. One of the key health-care issues and areas of progress for feminists in the U.S. has been the inclusion of women in medical research (Howes & Allina, 1994; Kass et al., 1996; Shumaker & Smith, 1994).

Adding Women to Medicine: Women's Health Research and Women Physicians

Joanne Howes and Amy Allina optimistically report that women now have a great deal of political influence over the priorities of health care research and policy. Politically, as women made inroads into Congress they began to push for the inclusion of women into existing medical research and for the re-focusing of research to address unique women's health issues. Women's health research is a viable political issue because the inequities were so flagrant (1994). Today health research along with heart disease, breast cancer, reproductive health, violence against women, nutrition and exercise are all issues receiving attention and the growing number of women in politics means that women's health issues will remain a political priority (Shumaker & Smith, 1994).

The growing attention to women's health in medical research has also been fueled by women's inroads into the medical profession. In 1970, only 7% of physicians were women: in 1994, there were over 100,000 female physicians (Howes & Allina, 1994). Women from all racial and ethnic backgrounds now make up to half or more of most medical school classes, but women and men are still segregated within areas of medical specialization (Lorber, 1997). And there is still little advancement to the top ranks of the medical profession for women. The progress that women have made must be viewed in light of the continuing obstacles facing women struggling for equal and humane health care (Macklin, 1993).

Continuing Disparities in Women's Health: Is Adding Women Enough?

While women have made progress in terms of making women's health part of the medical research agenda there are many remaining gaps and issues of concern from a sociological and feminist perspective. Paul Cotton notes women's progress in shaping the research agenda but cautions that there is still no input to the National Institutes of Health (NIH), one of the three federal agencies that has oversight for women's health research (Kass et al., 1996), from women's reproductive health care providers. And women's health care suffers in general from the continuing failure to include pregnant women or women of childbearing age in research trials (Cotton, 1993). Likewise, the health needs of women of color, especially as they relate to the experience of poverty, have been neglected because of the severe knowledge gaps that result from their exclusion in medical research (Lillie-Blanton et al., 1999). A great deal of medical research is funded by pharmaceutical companies and other industries that do not have to adhere to government guidelines for including women (Shumaker & Smith, 1994).

Sally Shumaker and Teresa Smith caution us that there are many remaining questions regarding the political efforts to address women's health priorities in medical research. It is crucial to consider: (1) who is establishing the research agenda, (2) who are the primary individuals designing and implementing the various studies, and (3) is this the best use of our health research dollars (1994: 192)? In general, feminists must question whether the fact that women's health has become a political agenda really ensures better care for women. Feminists are critical of the idea that simply bringing women into medicine and medical research, while retaining an overall biomedical framework, can transform the nature of medicine (Auerbach & Figert, 1995).

The biases of medical science remain; women's issues are still ignored and distorted. Many women in the U.S. today are still misdiagnosed or mistreated by medical professionals. One example is the frequency with which women with real medical complaints are disproportionately treated with psychotropic medications. Women in developing countries still lack basic human rights, including the right to health (Doyal, 1996; Joel, 1997). And these women are often subject to "brutish, humiliating treatment" at the "hands of the professional health care system" (Heise, 1993: 212). In general, women's reproductive health advocates have failed to acknowledge the reality of violence in many women's lives and have failed to acknowledge the fact that many third-world women need access to reproductive technologies over which they can exercise control (Heise, 1993). Here in the U.S. the reproductive needs of poor women have been increasingly ignored, in part because federal funding for family planning and reproductive services has all but disappeared (Blake, 1995–96).

Feminist sociologists caution that simply because there are more women in the ranks of medicine and politics fighting for (and defining) women's health issues does not mean that women's health will necessarily improve. And the lack of sociological input to current policy discussions may be the recipe for failure despite the fact that it was medical sociologists who helped to crystallize the current view of women's health as a gender issue:

> [M]edical and gender sociologists have documented empirically the gendered nature of health and medical care and the sex discrimination in the medical and scientific professions. Yet our critiques and data are not playing a more prominent role in the debate about women's health research policy (Auerbach & Figert, 1995: 126).

Auerbach and Figert note that already there is a backlash movement forming, both within the ranks of the NIH, and more generally, in the public press, against the inclusion of women into medical research (1995). Susan Bell writes that feminist theorists are now moving away from a strictly monolithic view of medicine and recognize many contradictions and ambiguities in the relationship between women and health. But ultimately medicine retains a great deal of power over women's lives and has many negative consequences (1995).

Feminist sociologists, and feminists studying the issue of medicalization from legal, criminological, public health, mental health and other disciplinary viewpoints, continue to document many ways in which medicine is related to and constructs complex notions about gender and gender inequality. They link this aspect of medicine's power as a social institution to complex ideas about the nature and activities of male and female bodies. They also note the important role of technology in determining the structure of medicine and the range and scope of medical bodily interventions. The changing structure of medicine is important, as is the related changing doctor-patient relationship. And finally, feminists continue to stress the way that medicine functions as a form of social control in women's lives.

A FEMINIST SOCIOLOGICAL CRITIQUE: ISSUES OF CHANGE AND CONTINUITY

Medicine, Gender and the Body: Women as Deviant

Karen Levy and other feminists insist that women's medical treatment is connected to very deeply-held ideas about the inferior and deviant nature of women's bodies. For Levy, the medical model of medicine creates an inherent conflict of interest between medical professionals and women:

> [C]onflict is embodied in the medical model, which defines women's natural functions as pathological, proposes surgery and drugs to treat problems, requires dependence on medical experts and maintains women's ignorance of their own bodies (1992: 49).

In general, medical services and professionals are not user-friendly for women (Doyal, 1996). Medical professionals are much more likely to perceive women's physical illnesses in terms of a psychological disorder: hence, the over prescription of psychotropic medications for women (Ashton, 1991). Paul Wise points out that child health advocates have virtually ignored the issue of women's health. Services for women start at the moment of conception and end with the woman's delivery (1995). Women's natural life conditions, especially those relating to a woman's reproductive capacity – birth control, pap smears & pelvic exams, menopause, pregnancy and childbirth – now involve women being treated as if they were somehow ill: "Women enter the medical system as patients, defined as ill and in need of care, no matter how normal their conditions are" (Levy, 1992: 45).

It is crucial to consider the role of the body – as the site of gendered meanings and practices – in an examination of women's relationship to medicine. For Sylvie Frigon, women have historically been seen as crazy and pathological, especially through the construction of their bodies as deviant. She stresses the continuity between the historical view of women as (deviant) witches and modern-day medical views of women. "As was the case for representations of the witches, psycho-biological and psychiatric interpretations also served and still serve to explain the deviance and criminality of women" (1995: 27).

Gender is one of the most significant factors in the transformation of physical bodies into social bodies. The bodily practices of gender start before birth and continue throughout our lifetime. Women and men engage in a variety of practices in order to transform their bodies into "markedly masculine and feminine" ideals. Many of these practices are related to medical issues and conditions of health and illness (Lorber, 1997: 3–4). In general, our society promotes a view of men's bodies as superior and ideal, while women's bodies are seen to be weak and to deviate from the male norm (Connell, 1995). Sports and medicine as institutions which promote images of health and illness are key in constructing ideas about gendered bodies and embodied social practices.

For instance, a recent book by Sharlene Hesse-Biber, poses that a "cult of thinness" is pervading American culture, fueled by the long standing patriarchal interest in women's bodies as decorative, and capitalism's interest in generating profit (1996). Jennifer Hargreaves sees a link between this cult and the growth of female sports and physical recreation in that women's bodies have increasingly become sexualized as sport is marketed specifically towards women (1992). Women's increasing participation in sports takes place within the context of the continuing commercialization of sports and leisure activities (Sage, 1990).

Clearly, the continuing increase in the numbers of women participating in sports since the passage of Title IX provides a challenge to definitions of women as the weaker sex, as well as empowering women as individuals (Sabo & Messner, 1993). Yet despite women's increasing presence, men still control the majority of sports organizations and professional sports remain the leading definer of masculinity in mass culture (Connell, 1995; Sage, 1990). Male and female professional athletes remain segregated by gender, and professional sports continue to generate the view that men's physical traits – speed, size, and strength – are the most desirable (Lorber, 1994). Women's participation in professional sports, particularly those considered masculine and requiring strength, such as body-building or boxing, is rife with contradictions and forces women athletes to construct images that are not sexually ambivalent for fear of being labeled lesbian. The most popular sports for women remain those that emphasize balance, flexibility, grace, and generally affirm a popular image of femininity (Hargreaves, 1994).

The media continue to focus on women athletes in sexualized, or, at best, ambivalent ways, and women still grace the cover of *Sports Illustrated* as swimsuit models (Cohen, 1993). One example of the ways in which women's bodies are sexualized through sports and mass media is found in the journalistic coverage of the recent 1999 Women's World Cup Soccer Games. Despite the athletic contributions of many U.S. team members to winning the final games, the media image that dominated afterwards is the picture of player Brandi Chastain removing her team shirt to reveal a sports bra.

Another example can be found in the growing popularity of aerobics for women. While this activity can certainly promote health and help empower women, the popularity of this form of exercise is also linked to the ways in which aerobics has been marketed to persuade women to participate in order to lose weight and improve their sex-appeal (Hargreaves, 1994). Similarly, young girls are taught through media exposure to resist exercise as it relates to health and instead prioritize fashion and body care: "Teenage femininity is thus coded in the language of consumerism and profit" (Hargreaves, 1994: 158). Young women increasingly view themselves as overweight and experience eating disorders. A sexualized version of health is being promoted – through sports, sportswear, women's magazines – and women have increasingly become obsessed with restructuring and enhancing their appearance through exercise, fashion, cosmetic surgery, and dieting (Hargreaves, 1992; Hesse-Biber, 1996).

The commercialization of women's bodies is linked to ideas about women's bodies as deviant and inferior to men's bodies. It also reflects a medical view of the human body that privileges the mind and sees the body as an appropriate cite for technological intervention. As Karen Levy states, the overall process of medicalization is dehumanizing:

> It objectifies people by treating them as machines in need of repair. The mind and spirit are separated from the body by an institution that is only interested in physiological processes. Through specialization, medicine focuses on increasingly smaller areas of the body, thus fragmenting care (1992: 11).

The sport sciences is a fairly recent and expanding field of medicine which exemplifies the ideological and material nexus of consumerism, technology, and the body as machine and object in need of repair. The goal of this flourishing industry – controlled mainly by men for men because professional sports remains a male bastion – is "performance enhancement" by any means possible. The body is treated as a machine and there is little concern for how the prescribed bodily interventions and manipulations interfere with the personal and social development of the athlete (Sage, 1990). A key feature in the development of sports medicine and other areas of medical specialization has been the growth of technology.

Medicine and Technology: The "Technological Imperative"

Karen Levy writes that one of the myths underlying the overall medicalization of society is the belief that technological medicine is necessarily *better* medicine (1992). Clearly, technology has come to play an increasingly important role in the medical enterprise, as the example of sports' medicine reveals. Sally Shumaker and Teresa Smith (1994) rely on Butter's concept of the "technological imperative" to explain the process by which childbirth has become firmly medicalized. This concept refers to the fact that Americans easily become dazzled by technology, and once it is introduced into social life, we want and expect it, no matter what the overall consequences. Further, Butter poses that because of this imperative, once technology has intervened into an arena of social life, ever-increasing amounts of technological innovation are demanded and desired by consumers, thus leading to a "cascade of interventions" (1993).

Carol Nadelson also notes the increasing cost and complexity of new technology and relates this trend to emerging ethical issues in the field of medicine. Increasing technology occurs within the context of limited resources for providing care. Thus medical professionals are now forced to make hard choices about who can have access to increasingly expensive and technologically advanced medical care. These dilemmas are partly the result of technological change but they are also related to two key changes in the organization of medicine – the changing doctor-patient relationship and the increasing corporate structure of medicine as an institution (1991).

Feminists have argued that the traditional structure of medicine, especially the configuration of the doctor-patient relationship, is key to the reproduction of gender inequality. The power imbalance and gender-stereotyped ways that doctors relate to patients have helped to reinforce men's power and strengthen sexist attitudes about women (Levy, 1992). In general there have been very few controls over the power of physicians in their interactions with patients and this has also been problematic for women (Doyal, 1996). But the shifting economic structure of medicine has impacted the nature of the doctor-patient relationship and the overall place of doctors in the medical hierarchy:

> A private entrepreneurial structure of care with considerable physician autonomy has been replaced by corporate management. Both patients and physicians are now managed. Like other practitioners, physicians have become hired hands. The medical system circumscribes practitioners' careers and care, and likely, their commitments. Corporate ownership has resulted in explicit measures to increase profit and to reduce costs (Charmaz & Paterniti, 1999: 2).

There is the possibility that the demise of the doctor-patient hierarchy may encourage more egalitarian relationships between health-care providers and patients. But because the transition in the management of health care is so current and yet widespread, it remains for feminist sociologists to conclusively assess its impact, and the impact of related changes, on gender relations and women's health. What is definitive at this point in time is that these changes – the increasing corporate structure of medicine, the changing role of the doctor, and, particularly, the growing importance and use of medical technology – will prove crucial in the future relationship between women's health and medicine. Feminists such as Susan Bell note that this relationship is full of "contradictions, ambiguities and uncertainties" and that medical science is "uncertain and ambiguous, infused with cultural assumptions about gender, and simultaneously productive of gendered meanings which can have negative consequences for women's lives" (1995: 469).

Bell argues that feminists have moved away from an "old social control model" of medicine. Yet, her own analysis reveals that in the case of the marketing of DES – diethylstilbestrol, the first synthetic estrogen – while physicians showed some resistance to the "diseasing of menopause," overall they saw it as a useful way to control the bodies of women. Bell argues that these physicians, despite their misgivings about the use of DES, embraced the "notion of the well-regulated human female" (1995: 43). Notwithstanding Bell's caution to resist a social control model of medicine, perhaps the most profound insight of feminist medical sociologists has been the range and depth of medicalization in women's lives.

Medicine as a Form of Social Control for Women

Karen Levy agrees with Conrad that medicalization can have positive effects. The medicalization of alcoholism, for instance, has optimistically recast the problem and provided the hope of treatment and cure when before there was only moral condemnation for the alcoholic. But, in general, she argues that the medicalization of women's natural life processes represents an expansion of social control. Because women's historical role has been the caretaker of the family, the medical establishment gains entry into the family and society by medicalizing women's lives. Plus, it is clear that women are the main consumers of health care, and that medicine has helped to shape and define the social roles available to women. Medical notions of women's frail, delicate, and emotionally unstable nature have been used to confine them to the home and hearth (1992).

While Conrad warns us about inferring a social control capacity to medicalization, he does note that "it is abundantly clear that women's natural life processes (especially concerning reproduction) are much more likely to be medicalized than men's and that gender is an important factor in understanding medicalization" (1992: 222). The breadth of medicalization in women's lives is staggering. A brief list of women's "medicalized" conditions includes childbirth (Doyal, 1996: Shumaker & Smith, 1994), pregnancy (Levy, 1992), menstruation (Frigon, 1996), birth control (Bell, 1995; Heise, 1993), infertility (Conrad, 1992), menopause (Frigon, 1996; Levy, 1992) and obesity (Conrad, 1992). Women's life experiences are more likely to be understood as syndromes, including pre-menstrual syndrome or PMS (Johnson & Kandrack, 1995), battered woman's syndrome or BWS (Frigon, 1995) and post-traumatic stress disorder or PTSD (Dobash & Dobash, 1992; Goodman et al., 1993).

Women undergo disproportionate surgical and pharmacological interventions in their lives (Ashton, 1991; Joel, 1997). Today, new surgical interventions – surgical breast implants and cosmetic surgery – have been used solely for aesthetic reasons (Davis, 1998; Rynbrandt & Kramer, 1995).

For Sylvie Frigon, there is no questioning the increased capacity for social control that results from the medicalization of women's lives. Medicine censures women today just as charges (and punishments) of witchcraft controlled women in the past. For Frigon, medicine is only part of the expanding system of social control that constructs and views women as deviant:

> Through prison, psychiatry, the institution of marriage, medicine, tranquilizers, electro-convulsive therapy, 'premenstrual syndrome', gynaecological surgery and laws (e.g. abortion and new reproductive technologies), society pathologizes and criminalizes women's bodies and women's minds (1995: 22).

In light of what appears to be an emerging debate about the consequences of medicalization, especially among feminist and other medical sociologists, I would like to turn to a few specific feminist analyses of the process. These analyses illustrate the role of technology, the drive for profits, and powerful notions about the body in their discussions of women's medicalization. They also shed light on the outcomes of medicalization for women and the future of this trend.

INSTANCES OF MEDICALIZATION

Childbirth

The medicalization of childbirth has been noted by many feminists and is related to the growth of medical technology, particularly the ability to monitor and sustain the lives of premature infants (Doyal, 1996; Levy, 1992). For Sally Shumaker and Teresa Smith the medicalization of childbirth and its construction as a pathological condition were also originally linked to Victorian era ideas about proper womanhood. Increasingly, births were attended only by doctors: family members were physically excluded and women were anesthetized and bodily removed from the birthing process. The result was that physicians gained control over the process and women's consignment to the delivery room was only a step away (1994).

What remains a puzzle is the extent to which feminists have been able to reverse the process. Peter Conrad proposes that childbirth has been demedicalized to some extent due to the efforts of feminists and consumer movements.

But feminists report that despite the efforts of activists who promote midwifery and researchers who report that out-of-hospital births are safer (cf. Tew, 1986) "medicalized childbirth remains firmly entrenched in this country" (Shumaker & Smith, 1994: 196). The practice of midwives, while enjoying a recent increase in this country, is still under continuous attack from the combined forces of the "medical establishment, hospital administrators, policymakers and ensurers" (194). Leslie Doyal reaches much the same conclusion noting that even in the U.S. where "pregnant women with resources can command considerable market power" childbirth remains firmly under medical control (1996: 52). Shumaker and Smith note the importance of the technological imperative in keeping childbirth medicalized, noting that women seem reluctant to choose alternatives even when they are available.

Cosmetic Surgery

Linda Rynbrandt and Ronald Kramer (1995) provide an analysis of the silicone breast implant case that highlights the ease with which this technology was introduced into the marketplace. They discuss the lack of Food and Drug Administration (FDA) attention to the problems associated with breast implants and stress that Dow Corning – the manufacturer – covered up these dangers by repeatedly faking reports and supplying the FDA with incomplete information. They stress the importance of the economic and cultural context of the introduction of this product:

> Structural forces in a capitalist, patriarchal society set the environment for organizations and individuals to inflict harm on consumers – in this case women – in the name of profit and growth. To compound the problem, cultural values that objectify and demean women also play an important role (222).

Kathy Davis writes more generally about the role of a market model of medicine, a consumer culture and powerful messages about the body in the rise of "the surgical fix" (1999). Davis notes that plastic surgery preceded cosmetic surgery and was first developed to help those who were disfigured by birth, by accident or by an illness. But these original technologies were quickly "borrowed" by cosmetic surgeons, and the field was aided by the growth of new technologies of its own, such as liposuction. In 1988, more than 2 million Americans had cosmetic surgery and over 90% of those operations were performed on women. Davis reminds us that what is missing from discussions about the industry's growth is the recipient. No one is asking why women are the primary patients or why they feel their bodies need to be altered in ways that put them at such physical risk.

While women remain the primary consumers of cosmetic or plastic surgery, the "huge financial potential in promoting body obsession and anxiety in men" (Hesse-Biber, 1996) is now being realized as men also increasingly turn to plastic surgery in search of the perfect body. In 1997, men spent almost $130 million on liposuction, face-lifts, nose reshaping and eyelid surgery, compared with $882 million spent by women on the same procedures. It is important to note that the majority of male patients resort to facial cosmetic surgery as they reach the later stages of life (cf. Wolfenden, 1997 and Hesse-Biber, 1996: 105). While older men are increasingly turning to surgery to enhance their appearance, there is far less gender symmetry in the situation present among the youngest generation. In 1998, approximately 1800 young women 18 and younger had their breasts enlarged, twice the number undergoing the same procedure in 1992. Baby boomers, men and women, make up over than 40% of the cosmetic surgery market today but technological advances, an expanding client pool, and the temporary nature of most surgical cosmetic enhancements or alterations may soon herald change in the average age of patients (Hayden et al., 1999).

Sharlene Hesse-Biber reports that the cosmetic surgery industry is now "a $5-billion-a-year industry" in which approximately 94% of the patients are women. She links this industry to the profitable fitness, diet and food industries as they all form part of our "cult of thinness" (1996). The diet industry profits from America's obsession with its weight to the tune of $30 billion annually and the fitness industry brings in over $40 billion a year. The growing emphasis on attractiveness is being amplified, in part, by the medical establishment, which links a trim smooth body, a full head of hair, and a dazzling smile to both health and happiness (Davidson, 1998; Hesse-Biber, 1996). Although it might not be accurate to relate all of these industries to the process of medicalization, it is clear that one of the motivating forces behind the "rise of the surgical fix" is the potential for profit in light of American's, and still predominantly women's, obsession with their bodily appearance

Fetal Abuse

Recently, attempts to prevent prenatal harm have led to court cases and legal sanctions against pregnant women. Joan Callahan and James Knight report that pregnant women are increasingly being incarcerated and forced to undergo medical and surgical treatments to prevent harm to their fetuses. They argue that a key factor in the process has been the development of prenatal therapies and technologies. It is the physicians who are developing these techniques who argue most vociferously about the rights of fetuses as patients (1992). Legally, this has led to an unprecedented challenge to women's autonomy in that the rights of the fetus are now superseding the rights of women (Nadelson, 1991). While women have historically been held accountable for child abuse, it is only recently that women have been legally subjected to child abuse statutes while pregnant (Stark & Flitcraft, 1994; Wise, 1995).

This historic shift is especially alarming in light of the fact that it is often poor women who lack access to health care and live with the reality of environmental, and other forms of, racism who wind up suffering under these punitive measures. In the case of drug-addicted mothers there are few modes of treatment available that would not either: (1) lead to punitive sanctions or (2) force them to get their child addicted to methadone. In 1992, approximately 60 women were prosecuted under state laws for taking drugs during their pregnancies and in some states, the babies were taken away and permanently placed in foster care. For Alexis Jetter and her colleagues, this current trend is linked to powerful notions about the health threat emanating from poor women's bodies:

> The maternal body has long been the container of social fears and fantasies. The HIV-positive, crack-addicted mother furnished a powerful amalgam of older imagery: the female body as dangerous and diseased; and poor women as immoral and irresponsible. Just as the welfare queen is seen as a 'cheater' who sucks dry scarce social resources, so the pregnant addict or addicted mother is seen as undermining the health of the social body (1995: 112).

Coerced Sterilization

The association of poor women's bodies with disease and immorality and the idea that they pose a threat to the public health of society is also what lies behind the rash of coercive policies that link sterilization procedures to women's welfare benefits. Despite the history of sterilization abuses against women of color (cf. Heise, 1993), Meredith Blake argues that we are now entering into the "beginning of a new era in state-sponsored reproductive control" (320). Technological advances in birth control have been key in this development. She highlights the role of the FDA and how it quickly approved Norplant for use by women despite many associated risks and side effects. Norplant has numerous side effects that are aggravated for poor women because of other health problems that they face, and black women are more likely to develop scar tissue that painfully interferes with Norplant's eventual removal. Since its FDA approval in 1990 there have been high removal rates and over 160 lawsuits have been filed against the manufacturers of Norplant–Wyeth–Ayerst – because of complications associated with its removal (1996).

Lori Heise, similar to Blake, notes that sterilization is not inherently bad and that many women choose to have this procedure done. She also points out that for many third-world women access to technologies such as Depo-Provera can help them to gain reproductive autonomy because it is a method that they can control. But as Blake notes, in the U.S., women are often not given the full picture when it comes to possible side effects. And many women on welfare are now forced to undergo these contraceptive procedures or lose a part, or all, of their welfare benefits: "Today's unfit appear to be poor, minority women who are being subjected to a form of state sponsored, coerced Darwinism" (341).

THE "MEDICO-LEGAL" ALLIANCE: THE FINAL STAGE IN WOMEN'S MEDICALIZATION

Kirsten Johnson and Mary-Anne Kandrack develop the concept of the "medico-legal" alliance in their analysis of the increasing use of PMS – pre-menstrual syndrome – as a legal defense for women. Noting that this defense is used primarily to argue insanity, especially in the cases of maternal-infant killings, they argue that the use of PMS as a legal defense is problematic for women. They attempt to explain why it is that women are so eager to accept PMS as a description of our experiences and our actions:

> Perhaps premenstrual discourse is popular because it offers women the discursive space in which to voice anger, frustration and depression in a culturally sanctioned manner. As well, PMS is accepted within the dominant culture precisely because it does not challenge, but rather (re)articulates stereotypical definitions of femininity (1995: 26).

Ultimately, the fact that PMS is used as a legal defense for women contributes to views of women as deviant, and entrenches the idea that women's deviance is located in their bodies, and is related to their capacity to reproduce. The intersection of legal discourse with psychiatric discourse represents the final stage in the medicalization of women's experiences (Kendall, 1991).

I propose that the legal punishment of women for prenatal abuse and the coercion of women on welfare to engage in sterilization procedures also represent examples of the medico-legal alliance. Elsewhere I have argued that the growth of the public health approach to violence against women is also part of the trend of medicalization in women's lives and that the growing medico-legal alliance must be considered in examining the usefulness of this development. What all of these instances of medicalization have in common is that they evidence the increasing role of the state in defining and utilizing medical definitions of women's behavior.

Peter Conrad argues that we cannot consider the growing expanse of public health campaigns to be part of the overall process of medicalization. He notes that during the 1980s, health promotion and wellness activities were touted as increasing individual health and reducing risk for disease but that this represents something distinct from the process of medicalization:

> [W]ith healthicization, behaviorial and social definitions are advanced for previously biomedically defined events. Medicalization proposes biomedical causes and interventions; healthicization proposes lifestyle and behaviorial causes and interventions. One turns the moral into the medical, the other turns health into the moral (1987).

But there is reason to believe that Conrad underestimates the ways in which public health campaigns represent the continuing medicalization of women's lives. Public health campaigns involve the power of the state and as such can be said to represent part of the "medico-legal" alliance referred to by Johnson and Kandrack. Further there is an important historical context to the implementation of public health campaigns that is relevant for women. Also, feminists note many ways in which the state contributes to violence against women. Finally, some feminists argue that a public health approach to medicine is supplanting the biomedical approach with potentially disastrous consequences for women.

Women as "Targets" of Public Health

Public health campaigns have historically targeted women and continue to do so. Women as mothers and as caretakers of the family have been the ones held responsible for their own health and the health of others (Nettleton, 1996). While women's health has historically been ignored, women – in their roles as wives and mothers – have always been the target of both covert and overt federal policies (Auerbach & Figert, 1995). Health promotion campaigns today routinely reproduce the gender stereotypes evident in commerical advertising. Women are portrayed either as passive, vulnerable and sexually available or as the guardians of the family's (and society's) well-being. As Deborah Lupton notes, public health media campaigns still vacillate between representing women as carers and women as the source of contagion, and encourage people to carry out surveillance on themselves and others (1999).

And while women are primarily the targets of public health campaigns, the sexist assumptions underlying this history are also evidenced by the failure to address women's danger of contracting HIV through heterosexual intercourse. Gene Corea stresses how the Centers for Disease Control and Prevention (CDC) classified the causes of a person's infection in a manner that made the spread of HIV among heterosexual women all but disappear, thus putting them in more danger. Transmission categories were set up hierarchically, so that the first risk behavior a person admitted to was determined to be the cause of transmission. This had the effect of distorting the frequency of heterosexual transmission because once a woman admitted illegal drug use that became the official cause. The CDC's campaigns to inform people of the dangers of transmission stereotypically focused on categories of people at risk rather than on behaviors that increased risk. And in doing so, the risk to women of contracting HIV through heterosexual intercourse was seriously downplayed (1999).

Corporate and Political Violence Against Women: The Role of the State

Feminists point to the negative consequences of public health campaigns for women. And they have also highlighted ways in which the state contributes to violence against women. For Susan Caulfield and Nancy Wonders, the state is ultimately responsible for violence against women in two ways. The state has directly participated in "crimes of commission" – direct acts of violence against women – but, more important to their argument, has also committed "crimes of ommission." These crimes relate to the state's active choice not to intervene or limit the serious harms directed toward women. They argue that the state has: (1) failed to create laws to address known harms against women, (2) has differentially applied existing laws, and has (3) allowed a culture of violence to permeate society with disastrous consequences for women's safety and health (1993).

Linda Rynbrandt and Ronald Kramer, in their analysis of the introduction of the silicone breast implant into the marketplace, argue that the failure of the state to regulate this product despite its known dangers is an example of a "crime of ommission." They also note that the state colluded with the manufacturers who are guilty of corporate violence against women. For them this is a clear-cut case of political and organizational crime involving the state and private industry (1995).

And Lori Heise points out the history of forced sterilization of women, citing this as the most notorious example of "medical violence" (212). In the 1970s many instances of involuntary sterilization of poor, minority, and mentally retarded women came to light. While the state has supposedly ended these more flagrant abuses of women's right to bodily self-determination, the logic behind coercive welfare policies of today draws on the same arguments and rationales used to justify these earlier abuses (1993). Although Peter Conrad draws a sharp distinction between medical and moral views on the causes of health and illness, the reality is that women have always been held responsible for the moral and physical health of themselves and others. Thus, public health campaigns that increasingly link physical health to individual morality and continue to target women implicitly blame women for their own and for other's immorality and illness.

THE POWER OF MEDICINE TO CONTROL WOMEN'S LIVES: CLOSING THOUGHTS

The "medico-legal" alliance is a reality in women's lives. Medicalized conditions are now becoming part of the legal landscape and the power of the law is used to force women to submit to medical treatments. For Sylvie Frigon women's depiction in public health campaigns represents just one aspect of the overall way in which women's bodies are continuously portrayed as being in danger and dangerous at the same time. Medicine as an institution is only one facet of what Frigon proposes is a "disciplinary matrix" of control that serves to neutralize women's political protest. This matrix consists of the criminalization, sexualization and medicalization of women's bodies and minds (1995). Frigon has analyzed the use of PMS, and BWS – battered women's syndrome – in the legal defense of women (1996). She notes that, while the analysis supporting the reality of PMS is biological and medical, and the evidence underlying the existence of BWS is psychological and sociological, "the question of the unruly bodies and minds of women is always central" (95). The legal recognition of both as conditions affecting women serves to "translate women's behavior as an individual disorder/pathology" (97).

For Sarah Nettleton, the distinction that Conrad draws between medicine and public health is fast dissolving. She argues that we are witnessing the evolution of biomedical science into the NPHM – a new paradigm of health and medicine that centers on key notions of "risk," "surveillance," and the "rational self." The core policy dimension of this new model is health promotion, aided by technological advances which make the detection of risk factors more and more sophisticated. Health care is increasingly becoming concerned with the management of "risky" groups and behaviors, and traditional exhortations to women to protect the health of others, through new modes of surveillance and management of risk, will likely continue. She predicts that the social and material circumstances of a "healthy" lifestyle will increasingly be obscured and forgotten. Women will increasingly be held accountable for prenatal harm as their experience is examined only in terms of the risk factors that they engage in while pregnant. And the "tyranny of a 'healthy' body is likely to become more evident" as images of health increasingly become sexualized and individualized (1996).

The need for feminist sociologists to be involved in policy debates about women's and men's health issues is paramount. It is clear that there are many crucial links between the medical and legal system as they interact and shape women's (and men's) daily lives. The feminist analyses that I have drawn from in this essay stress that medicalization, while occuring on a society-wide level, has unique implications for women. The complex ways that various medical fields – psychology, psychiatry, biomedical science – interact and contribute to women's social control need to be explored. And the implications of the technological and organizational changes that are occurring, and whether or not these changes herald a new paradigm of medicine, are questions that need to be addressed. In light of the way in which women's and children's social issues are stripped of their connections by medical professionals (and academics), there is much work to be done weaving back together the shared elements of their experiences. It is clear that children's lives also continue to be medicalized, and that this process is linked to their increasing social control (cf. Sutton, 1994). How these processes and outcomes for children are connected to women's medicalization needs to be explored.

Finally, it is clear that we need to better examine the process of medicalization from the point of view of the women who are experiencing it. Women are not merely passive objects of medical control. While they have often been coerced into medical treatment, there are also many ways that women can resist, encourage or embrace medicalization, especially as they have historically been the caretakers of other's health. As Susan Bell writes, feminists have revealed many tensions and contradictions between women and medicine, and have discovered positive and negative consequences to this complex relationship. But while we unearth and disentangle these complexities, technological advances are also rapidly occurring and the potential for women's regulation continues to expand over decades (1995). The need for a feminist medical sociology that can inform policy debates and enrich our understanding of the dynamics and consequences of medicalization for society has never been greater.

ACKNOWLEDGMENTS

I would like to offer my thanks to Dorie Klein, Jennie Jacobs Kronenfeld, and especially to Kathleen Montgomery, for their helpful comments on earlier drafts of this chapter. I would also like to acknowledge the on-going support of the Robert Presley Center for Crime and Justice Studies at the University of California—Riverside, and the encouragement and support of its Director, Robert Nash Parker. An earlier draft of this article was presented at the American Sociological Association's 1999 Conference in Chicago.

NOTES

1. For a summary of the contributions of medical sociologists see the article "Women's Health Research: Public Policy and Sociology" in the *Journal of Health and Social Behavior* (1995, extra issue, pp. 115–132) by Judith Auerbach and Anne E. Figert.

2. Susan Bell (1995) lists several feminist publications as key to the early women's health movement. These works include *The Doctor's Case against the Pill* by Barbara Seaman (1969) and *Witches, Midwives and Nurses: A History of Women Healers* by Barbara Enhrenreich and Diedre English (1973). Joanne Howes and Amy Allina (1994) also credit the publication of the first Boston Women's Health Book Collective entitled *Our Bodies Ourselves* in 1970 with ushering in a new era in women's health.

REFERENCES

Ashton, H. (1991). Psychotropic drug prescribing in women. *The British Journal of Psychiatry, 158*, 30–35 (Supplement. 10, edited by The Women and Mental Health International Conference Committee).

Auerbach, J. D., & A. E. Figert. (1995). Women's Health Research: Public Policy and Sociology. *Journal of Health and Social Behavior*, (extra issue), 115–132.

Bell, S. E. (1995). Gendered Medical Science: Producing a Drug for Women. *Feminist Studies, 21*(3), 469–500.

Blake, M. (1996). Welfare and Coerced Contraception: Morality Implications of State Sponsored Reproductive Control. *University of Louisville Journal of Family Law, 34*(2), 311–344.

Boston Women's Health Collective. (1970). *Our Bodies Ourselves*. New York: Simon and Schuster.

Butter, I. H. (1993). Premature Adoption and Routinization of Medical Technology: Illustrations from Childbirth Technology. *Journal of Social Issues, 49*(2), 11–34.

Callahan, J. C., & Knight, J. W. (1992). Prenatal Harm as Child Abuse? *Women and Criminal Justice, 3*(2), 5–33.

Caulfield, S. L., & Wonders, N. A. (1993). Personal AND Political: Violence against Women and the Role of the State. In: K. Tunnell (Ed.), *Political Crime in Contemporary America: A Critical Approach* (pp. 79–100). New York: Garland Press.

Charmaz, K., & Paterniti, D. A. (1999). Introduction: Health, Illness and Healing: Society, Social Context, & Self—An Anthology. In: K. Charmaz & D. A. Paterniti (Eds), *Health, Illness and Healing: Society, Social Context, & Self—An Anthology* (pp. 1–3). Los Angeles: Roxbury.

Cohen, G. L. (1993). Media Portrayal of the Female Athlete. In: G. L. Cohen (Ed.), *Women in Sport: Issues and Controversies* (pp. 171–184). Newbury Park, CA: Sage.

Connell, R. W. (1995). *Masculinities*. Los Angeles: University of California Press.

Conrad, P. (1992). Medicalization and Social Control. *Annual Review of Sociology, 18*, 209–232.

Conrad, P. (1975). The Discovery of Hyperkinesis: Notes on the Medicalization of Deviant Behavior. *Social Problems, 23*, 12–21.

Conrad, P., & Schneider, J. (1980). *Deviance and Medicalization: From Badness to Sickness*. St. Louis: Mosby.

Corea, G. (1999). The Invisible Epidemic: The Story of Women and AIDS. In: K. Charmaz & D. A. Paterniti (Eds), *Health, Illness and Healing: Society, Social Context, & Self—An Anthology* (pp. 463–466). Los Angeles: Roxbury.

Cornwell, J. (1984). *Hard-Earned Lives: Accounts of Health and Illness in East London*. New York: Tavistock.

Cotton, P. (1993). Women Physicians Target Barriers (Medical News & Perspectives). *Journal of the American Medical Association*, *269*(8), 965.

Davis, K. (1999). The Rise of the Surgical Fix. In: K. Charmaz & D. A. Paterniti (Eds), *Health, Illness and Healing: Society, Social Context, and Self—An Anthology* (pp. 302–321). Los Angeles: Roxbury.

Davidson, D. K. (1998). Marketing this 'hope' sells our profession short. *Marketing News*, *32*(15), 6.

Dobash, R. E., & Dobash, R. P. (1992). *Women, Violence and Social Change*. New York: Routledge.

Doyal, L. (1996). The Politics of Women's Health: Setting A Global Agenda. *International Journal of Health Services*, *26*(1), 47–65.

Ehrenreich, B., & English, D. (1973). *Witches, Midwives and Nurses: A History of Women Healers* (Glass Mountain Pamphlet No. 1). New York: Feminist Press.

Foucault, M. (1973). *The Birth of the Clinic*. New York: Vintage.

Foucault, M. (1965). *Madness and Civilization*. New York: Random.

Fraser, L. (1999). The Hard Body Sell. *Mother Jones*, *24*(2), 31.

Frigon, S. (1996). A Gallery of Portraits: Women and the Embodiment of Difference, Deviance, & Resistance. In: T. O'Reilly-Fleming (Ed.), *Post-Critical Criminology* (pp. 78–110). Scarborough, Ontario: Prentice-Hall Canada Inc.

Frigon, S. (1995). A Genealogy of Women's Madness. In: R. E. Dobash, R. P. Dobash, & L. Noaks (Eds), *Gender and Crime* (pp. 20–48). Cardiff: University of Wales Press.

Goodman, L. A., Koss, M. P., Fitzgerald, L. F., Russo, N. F., & Keita, G. P. (1993). Male Violence Against Women: Current Research and Future Directions. *American Psychologist*, *48*(10), 1054–1058.

Hargreaves, J. (1994). *Sporting Females: Critical Issues in the History and Sociology of Women's Sports*. New York: Routledge.

Hargreaves, J. (1992). Sex, Gender and the Body in Sport and Leisure: Has There Been a Civilizing Process? In: E. Dunning & C. Rojek (Eds), *Sports and Leisure in the Civilizing Process: Critique and Counter-Critique* (pp. 161–182). Buffalo: University of Toronto Press.

Hayden, T., Figueroa, A., Downey, S., Sieder, J. J., Laverman, J., Pierce, E., & Roberts, E. (1999). Our Quest to be Perfect: You don't have to be wrinkled or rich to go under the knife. New techniques are luring more and younger patients, while doctors hurry to set up shop. But there are risks as well as options in cosmetic surgery's new age. *Newsweek, 52*.

Heise, L. L. (1993). Reproductive Freedom and Violence against Women: Where are the Intersections? *The Journal of Law, Medicine & Ethics*, *21*(2), 206–216.

Hesse-Biber, S. J. (1996). *Am I Thin Enough Yet? The Cult of Thinness and the Commercialization of Identity*. New York: Oxford University Press.

Howes, J., & Allina, A. (1994). Women's Health Movements. *Social Policy*, *24*(4), 6–16.

Illich, I. (1976). *Medical Nemesis*. New York: Pantheon.

Jetter, A., Orleck, A., & Taylor, D. (1995). Mothers at Risk: The War on Poor Women and Children. In: G. J. Demko & M. C. Jackson (Eds), *Populations at Risk in America:*

Vulnerable Groups at the End of the Twentieth Century (pp. 104–128). Boulder, CO: Westview Press.

Joel, L. A. (1997). Women's Health as a Human Rights Issue. *The American Journal of Nursing, 97*(4), 7.

Johnson, K. K., & Kandrack, M. A. (1995). On the Medico-Legal Appropriation of Menstrual Discourse: The Syndromization of Women's Experiences. *Resources for Feminist Research, 24*(1), 23–27.

Kass, N. E., Taylor, H. A., & King, P. A. (1996). Harms of Excluding Pregnant Women from Clinical Research: The Case of HIV-Infected Pregnant Women. *Journal of Law, Medicine & Ethics, 24*, 36–46.

Levy, K. B. (1992). *The Politics of Women's Health Care: Medicalization as a Form of Social Control*. Las Colinas, Texas: Ide House.

Lillie-Blanton, M., Martinez, R. M., Taylor, A. K., & Robinson, B. G. (1999). Latina and African American Women: Continuing Disparities in Health. In: K. Charmaz and D. A. Paterniti (Eds), *Health, Illness and Healing: Society, Social Context, & Self—An Anthology* (pp. 395–414). Los Angeles: Roxbury.

Lorber, J. (1997). *Gender and the Social Construction of Illness*. Thousand Oaks, CA: Sage.

Lorber, J. (1994. *Paradoxes of Gender*. New Haven, CT: Yale University Press.

Lupton, D. (1999). The Imperative of Health: Public Health and the Regulated Body. In: K. Charmaz & D. A. Paterniti (Eds), *Health, Illness and Healing: Society, Social Context, and Self—An Anthology* (pp. 42–47). Los Angeles: Roxbury.

Macklin, R. (1993). Women's Health: An Ethical Perspective. *The Journal of Law, Medicine & Ethics, 21*(1), 23–29.

Messner, M. A., & Sabo, D. F. (1993). Whose Body is This? Women's Sports and Sexual Politics. In: G. L. Cohen (Ed.), *Women in Sport: Issues and Controversies* (pp. 15–26). Newbury Park, CA: Sage.

Mishler, E. G. (1981). Viewpoint: Critical Perspectives on the Biomedical Model. In: Mishler et al. (Eds), *Social Contexts of Health, Illness, and Patient Care* (pp. 1–23). Cambridge, UK: Cambridge University Press.

Nadelson, C. C. (1991). Emerging Issues in Medical Ethics. *The British Journal of Psychiatry, 158*, 9–16 (Supplement. 10, edited by The Women and Mental Health International Conference Committee).

Nettleton, S. (1996). Women and the New Paradigm of Health and Medicine. *Critical Social Policy, 16*(3), 33–53.

Pfohl, S. J. (1977). The 'Discovery' of Child Abuse. *Social Problems, 24*, 310–323.

Rynbrandt, L. J. (1995). Hybrid Nonwomen and Corporate Violence: The Silicone Breast Implant Case. *Violence Against Women, 1*(3), 206–227.

Sage, G. H. (1990). *Power and Ideology in American Sport: A Critical Perspective*. Champaign, IL: Human Kinetics Books.

Seaman, B. (1969). *The Doctor's Case against the Pill*. New York: Doubleday.

Shumaker, S. A., & Smith, T. R. (1994). The Politics of Women's Health. *Journal of Social Issues, 50*(4), 189–202.

Stark, E., & Flitcraft, A. H. (1994). Women and Children at Risk: A Feminist Perspective on Child Abuse. In: N. Krieger & E. Fee (Eds), *Women's Health, Politics, and Power: Essays on Sex/Gender, Medicine, and Public Health* (pp. 307–332). Amityville, NY: Baywood Publishing Company.

Sutton. J. R. (1994). Children in the Therapeutic State: Lessons for the Sociology of Deviance and Social Control. In: G. S. Bridges & M. A. Myers (Eds), *Inequality, Crime & Social Control* (pp. 227–248). San Francisco: Westview Press.

Tew, M. (1986). Do Obstetric Intranatal Interventions Make Birth Safer? *British Journal of Obstetrics and Gynaecology, 93*(7), 659–674.

Wise, P. (1995). Child Beauty, Child Rights and the Devaluation of Women. *Health and Human Rights* (Special Focus: Women's Health and Human Rights), *1*(4), 472–476.

Wolfenden, W. J. Jr. (1997). A Lean, Mean Surgery Machine: A cosmetic surgeon describes his practice, which is designed to make people look younger and feel better about themselves. *Forbes, 16*, S24–25.

PART II

SOCIOECONOMIC STATUS AND RACE IN HEALTH AND HEALTH CARE

CARE-SEEKING PATTERN AND OUTCOME OF HOSPITALIZED PNEUMONIA PATIENTS WHO ARE SOCIOECONOMICALLY DISADVANTAGED

Jay J. Shen, Jonathan B. Perlin, Yi-Hua Xu and Thomas T. T. Wan

INTRODUCTION

Socioeconomic status (SES) has long been a focus of health policy and health services research in examining variation in access to and outcomes of health services across different socioeconomic populations (Gold, 1995). For decades, access to care and related outcomes of health services of both economically and culturally disadvantaged populations have continued being important issues in health policy and health care delivery. Socioeconomic status remains a major predictor of variations in health care outcomes (Moore, 1997; Williams & Collins, 1995). Concern has been promoted by persistent and widening gaps in access and outcome of major diseases between disadvantaged persons and those who are more privileged in society (Altman, Reinhardt & Shields, 1998; Lillie-Blanton, 1999; Nickens, 1995; Williams & Collins, 1995).

This study emphasizes pneumonia, the number one cause of death from infectious diseases and the sixth leading cause of death in the United States, whose relationships with socioeconomic status has been relatively less studied.

Health, Illness, and Use of Care: The Impact of Social Factors, Volume 18, pages 97–120.

ISBN: 0-7623-0740-4

Over four million cases of pneumonia are estimated to occur each year, with about 1.1 millions hospitalizations (American Thoracic Society, 1993; Garibaldi, 1985; Niederman et al., 1998). The aggregated costs of the hospitalization are estimated from $4 billion to $7.5 billion per year (Fine et al., 1997; La Force, 1985; Niederman et al., 1998). Pneumonia patients who require hospitalization experienced high risk of mortality, at an average rate of around 14%, with the highest rate approaching 25% to 37% for patients admitted to intensive care units (American Thoracic Society, 1993; Fine et al., 1996).

Socioeconomic status, widely used as a proxy for social class, generally is measured by race, education, income, insurance status, or occupation (Mueller, Patil & Boilesen, 1998; Nicksens, 1995; Williams & Collins, 1995). Evidence on socioeconomic status as a barrier in access to health care and as a predictor of outcome has continued to emerge (Lillie-Blanton, 1999; Schoen & DesRoches, 2000; Smith & Chang, 1995). Although education is a robust indicator of SES (Schulman et al., 1995; Williams & Collins, 1995) and income level reflects financial capability of getting access to health services (Mueller, Patil & Boilesen, 1998), both are usually obtained from specific surveys and often not available from administrative data repositories (Schulman et al., 1995). As with education and income, information on occupation is rarely available.

Insurance status and race are more frequently used socioeconomic measures in research. Insurance status, similar to income and as a measure of economic inequality, is often used to indicate the degree of financial barrier preventing access to health services (Gold, 1998; Mueller, Patil & Boilesen, 1998). It continues to be a remarkably robust determinant of variation in outcomes such as mortality and health status (Berk & Schur, 1998; Schoen & DesRoches, 2000; Young & Cohen, 1991). In some cases, economic access to health care becomes a more dominant factor than clinical factors to affect the treatment and outcome of patients get (Moore, 1997).

Using race as a socioeconomic measure appears more complicated. Despite being used to reflect cultural barriers to health care (Belgrave, Wykle & Choi, 1993; Mueller, Patil & Boilesen, 1998), race can incorporate biological, social, and cultural characteristics of patients and can refer to both genetic and behavioral traits. Many studies use race as a proxy for other socioeconomic factors that are not currently available in health services research leading to disparate conclusions (Schulman et al., 1995; Williams & Collins, 1995). For example, the question of whether African-Americans incur worse health care outcomes on the basis of socioeconomic status than do non-Hispanic whites, the literature so far has demonstrated mixed results (Lillie-Blanton, Parsons,

Gayle et al., 1996; Williams & Collins, 1995). This inconsistency often results from the lack of control for socioeconomic factors other than race because evidence has shown that racial differentials in outcome become less significant when data are stratified or adjusted for other socioeconomic attributes (Lillie-Blanton, Parsons, Gayle et al., 1996; Schulman et al., 1995; Williams & Collins, 1995). Hence, race can become a more valid exploratory measure of socioeconomic status if it is used with other socioeconomic measures (Schulman et al., 1995).

Another less developed line of inquiry related to socioeconomic status and outcome is the multi-dimensional vulnerability scenario in which the effects on persons' health care outcomes are examined if they bear two or more socioeconomic risk factors. Multi-dimensional vulnerability scenarios demonstrate increased disadvantages for both obtaining health services and for having good clinical outcomes (Belgrave, Wykle & Choi, 1993). One type of the multi-dimensional vulnerability, "double-jeopardy, " depicts in socioeconomic status, such as being both minority and poor. The double-jeopardy hypothesis originally proposed that African-American elderly face more of a burden than non-Hispanic white elderly do (Belgrave, Wykle & Choi, 1993), and is generalized to examine other socioeconomic scenarios related to the double jeopardy (Rich, Rich & Mullins, 1995; Salmon, 1994).

In addition to individual-based socioeconomic status, the community-based socioecological characteristics are important factors affecting health, too. It has been theorized that outcome differences among populations can be explained largely by community discrepancy in sanitation, nutrition, housing, and other environmental elements which are largely determined by income differences (Bloomberg, Meyers & Braverman, 1994). The socioecological perspective is particularly important when potential variations in disease burden associated with different populations belonging to different socioeconomic classes are explored. Combining both individual-based socioeconomic factors and community-based socioecological factors can elucidate relationships between socioeconomic status and outcome more clearly.

Although studies of Medicare patients by Ayanian and colleagues (1999) found that African-American patients with pneumonia received lower quality of care than other patients, little has been done to explore the relationship between socioeconomic status and the outcome of pneumonia, especially for pneumonia patients who carry more than one socioeconomic risk factor. Also, although the association of lower socioeconomic status with more advanced disease at presentation, which strongly affects outcome, has been observed by studies focusing on other diseases (Haywood, Ell, deGuman et al., 1993), no studies focusing on pneumonia have been found in literature.

Limited knowledge exists regarding the important relationship between socioeconomic status and pneumonia, particularly more serious pneumonia requiring hospital admission. This study addresses multiple dimensions of socioeconomic disadvantages, including the concept of double-jeopardy. Incorporating both socioeconomic and socioecological information and using a national dataset, this study explores associations of socioeconomic status with outcome, as well as care-seeking patterns of patients requiring hospitalization.

METHODS

Data and Sample

Data were abstracted from the National Inpatient Sample (NIS), Release 3, 1994, collected by the Healthcare Cost and Utilization Project (HCUP-3), the Agency for Healthcare Research and Quality (AHRQ) (formerly the Agency for Health Care Policy and Research (AHCPR)). The dataset contains approximately 20% of the total inpatient discharges in 1994, drawn as a stratified randomized sample from hospitals in 17 states that participated in the HCUP. Pneumonia discharges were identified based on the International Classification of Disease, Ninth Edition, Clinical Modification (ICD-9-CM) codes used by the Medicare Quality Indicator System (MQIS). Pneumonia discharges were those who had a principal discharge ICD-9-CM codes 480.0–480.9, 481, 482.0–482.9, 483.0–483.8, 485, 486, 487.0, 507.0 or who had a principal discharge diagnosis of respiratory failure (ICD-9-CM code 518.81) and a secondary diagnosis of pneumonia (Meehan et al., 1997).

The unit of analysis was hospital discharge. Discharges in six states, however, were excluded. Information about expected payers for patients' care was not reliable in hospitals of Iowa. Race information was not available in Arizona, Illinois, Oregon, Pennsylvania, and Washington. Therefore, among a total of 234,725 pneumonia discharges in the 17 states, 172,981 in the remaining 11 states were extracted as the sample for the study. The 11 states were California, Colorado, Connecticut, Florida, Kansas, Massachusetts, Maryland, New Jersey, New York, South Carolina, and Wisconsin.

Measures

According to information contained in the NIS data, three dependent variables were selected for this study. First, admission through emergency room (ER) was used to measure the care-seeking pattern of pneumonia patients. Timely treatment of pneumonia makes it less likely to require ER admission. Some studies reported that race, health insurance coverage, and difficulty obtaining transportation to a physician's office were associated with more ER use (Agency for Health Care Policy and Research, 1998; Baker, Stevens & Brook, 1996; White-Means, 1995). Second, existence of comorbidity conditions was examined along with the admission pattern. Clinical evidence indicates that patients having comorbidities are more likely than those without comrobidities to be admitted through emergency room (Boersma, 1999). It was expected, hence, that the ER admission is associated with a higher probability of having comorbidities.

The third dependent variable, also the most important one – in-hospital mortality – was used as an outcome measure. It was adjusted for patient's age, sex, and comorbidities that are potentially clinical risk factors for hospital death. Of the 30 comorbidities, identified by Elixhauser and colleagues (1998), peripheral vascular disorder was not found among all discharges. The remaining 29 comorbidities included in the analysis were congestive heart failure, cardiac arrhythmias, valvular disease, pulmonary circulation disorders, hypertension, paralysis, other neurological disorders, chronic pulmonary disease, diabetes, hypthyroidism, renal failure, liver disease, peptic ulcer disease excluding bleeding, AIDS, lymphoma, metastatic cancer, solid tumor without metastatsis, rheumatoid arthristis/collagen vascular diseases, coagulopathy, obesity, weight loss, fluid and electrolyte disorders, blood loss anemia, deficiency anemias, alcohol abuse, drug abuse, psychoses, and depression. A preliminary analysis indicated that four comorbidities, uncomplicated diabetes, rheumatoid arthristis/collagen vascular diseases, blood loss anemia, and alcohol abuse were not significantly associated with mortality while drug abuse and psychoses showed a marginal association with mortality; therefore, the four conditions were not included in the risk-adjustment model.

The method developed by Elixhauser and colleagues improves the measurement of comorbidities in large administrative datasets in three ways: (a) it adds more comorbidities to the Charlson morbidity list (Charlson et al., 1987; Deyo et al., 1992) and drops several that were never related to outcomes or that were conceptually inappropriate; therefore, it allows for a flexible approach to measuring comorbidities; (b) it uses the DRG screening to eliminates comorbidities that are related to the principal diagnosis; and (c) based on complications described by DesHarnais (1990), it eliminates those conditions that were almost sure to be complications rather than comorbidities (Elixhauser et al., 1998).

Socioeconomic status, as the explanatory variable, was measured by health insurance status, race, and median income by zip code of residence. Insurance status, regarded as the patient's mechanism of payment to obtain access to entering the health services system (Gold, 1998), was grouped as Medicare, Medicaid, private insurance including HMO/prepaid health plans, and uninsured. Race, categorized as non-Hispanic White, African-American, and other races, was used as a proxy for potential cultural differences between ethnic groups as well as other unidentified socioeconomic traits (Belgrave, Wykle & Choi, 1993; Mueller, Patil & Boilesen, 1998; Schulman et al., 1995; Williams & Collins, 1995). Due to a "universal" coverage, Medicare patients were examined both together with and separately from patients with other insurance status.

The median income by zip code of patient residence was used to measure the community's overall socioecological conditions that have been shown associations with health care outcome (Claudio, Tulton, Doucette et al., 1999; Pappas, Hadden, Kozak et al., 1997; Smith, Wentworth, Neaton et al., 1996). Median income by zip code ranged from less than $15,000 to over $45,000 constituting two strata, and income was also stratified at three intermediate levels.

The preliminary analysis revealed much stronger correlations of race and insurance status than that of the median income by zip code to in-hospital mortality. Race and insurance status, thus, were selected to construct double-jeopardy variables. Patients in double jeopardy were defined as those who had both inferior health insurance and vulnerable race. The inferior insurance status was defined as either Medicaid or uninsured, as opposed to the better insurance status – private insurance. The vulnerable race was defined as being African-American as opposed to non-Hispanic whites who are less vulnerable in seeking health services. Generalized from Ferraro's description (1989), if the double-jeopardy hypothesis holds, first, patients with inferior insurance status should have higher pneumonia mortality than should the privately insured patients; and second, mortality variance between patients of vulnerable race and patients of non-vulnerable race should be greater for patients having inferior insurance status than for those in private insurance programs. Hence, both insurance status and race, as well as their interaction term, were included in a multivariate analysis in order to empirically observe any salient double jeopardy phenomenon.

Further, in order to examine the magnitude of difference in outcome between those who carried two disadvantageous socioeconomic characteristics and those who had two favorable ones, two subgroups of discharges were identified. The "extremely unfavorable group" was composed of those who were covered by Medicaid or were uninsured, and who were also African-American. The "extremely favorable group" consisted of those who were covered by private insurance and who were also non-Hispanic white. The analysis defined a dummy variable (equal to 1 if a discharge belonged to the extremely unfavorable group and equal to 0 if the discharge belonged to the extremely favorable group). Wider variations were expected between the two groups in mortality as well as in ER admission and existence of comorbidity conditions. In addition, length of stay (LOS) and the total charges per discharge, two measures as the proxy of resource uses or efficiency, were also compared.

Finally, since the NIS is a multi-state data source, 10 dummy variables for 10 states, not including California, used as the reference, were created to control for geographic differences that might be related to pneumonia patients' admission pattern and outcome.

Analytical Techniques

The SAS software (SAS Institute, 1990, 1997) was used to conduct logistic regression in order to predict the adjusted mortality odds as well as the ER admission odds and comorbidity existence odds for each pneumonia patient and to detect salient SES attributes. In addition, the general linear model (GLM) (SAS Institute, 1990) was used to compare length-of-stay and total charges per discharge between the "extremely favorable" and the "extremely unfavorable" groups. Due to the skews of the data distributions, a natural logarithm transformation was performed for all dependent variables, which were continuous, in the GLM model.

RESULTS

Patient SES and Hospitalization Characteristics: Medicare, Non-Medicare, and Overall

Descriptive characteristics of patients' socioeconomic status and hospitalization for all pneumonia discharges, as well as for Medicare and for non-Medicare discharges are summarized in Table 1. The greatest difference in discharge characteristics was observed between Medicare and non-Medicare patients. Not surprisingly, the non-Medicare group (average 32.5 years old) was much younger that the Medicare group (average 77.3 years old). The Medicare group had a higher percentage of females, 51.5% while the percentage of females in the non-Medicare group was about 47.1%. While the Medicare patients were composed of 14.3% of minorities the non-Medicare patients had 38.7%. About 59.0% of the Medicare patients were admitted from the emergency room and 95.1% of the patients bore at least one comorbidity.

Table 1. . Demographic and Hospitalizational Characteristics of Patients

	All Patients		Medicare Patients		Non-Medicare Patients	
	Frequency	Percent (%)	Frequency	Percent (%)	Frequency	Percent (%)
Sex						
Male	86,996	50.3	49,647	48.5	37,349	52.9
Female	85,977	49.7	52,689	51.5	33,288	47.1
Race						
Non-Hispanic White	128,630	78.0	86,183	85.7	42447.0	61.3
African American	24,588	14.9	8,302	8.3	16,286	23.5
Others	11,663	7.1	6,079	6.0	10,505	15.2
Insurance status						
Medicare	102,339	59.2	102,339	100.0		
Medicaid	26,453	15.3			26,453	37.5
Private Insurance	35,899	20.8			35,899	50.9
Uninsured	8,192	4.7			8,192	11.6
Median income by zip code ($)						
< 15,000	5,366	3.5	2,250	2.5	3,116	5.0
–25,000	45,034	29.4	25,993	28.5	19,047	30.6
–35,000	59,066	38.5	35,767	39.3	23,299	37.4
–45,000	27,914	18.2	17,038	18.7	10,876	17.4
> 45,000	16,046	10.5	10,057	11.0	5,989	9.6
ER Admission						
Yes	96,989	57.2	59,284	59.0	37,705	54.5
No	72,590	42.8	41,156	41.0	31,434	45.5
Existence of Comorbidity						
Yes	145,090	83.9	97,295	95.1	47,795	67.7
No	27,891	16.1	5,044	4.9	22,847	32.3
Discharge status						
Discharged alive	157,288	91.0	89,446	87.5	67,842	96.1
Discharged dead	15,564	9.0	12,781	12.5	2,783	3.9
	Mean ($n = 172,793$)	St.d.	Mean ($n = 102,334$)	St.d.	Mean ($n = 70,456$)	St.d.
Age	59.0	29.2	77.3	11.6	32.5	26.5
Length of stay	7.7	6.6	8.8	6.9	6.0	5.8
Total charges ($)	12,865.0	17516.1	14,093.6	16,982.4	11,082.8	18,115.2

St.d. = Standard deviation.

In contrast, only 54.5% of the non-Medicare patients were admitted from ER and 67.7% of the patients had a comorbidity. The average unadjusted hospital mortality rates were 12.5% for Medicare patients and 3.9% for non-Medicare patients. The average length-of-stay for Medicare patients (8.8 days) was about three days longer than that for the non-Medicare patients (6.0 days). Finally, non-Medicare patients incurred a charge of $14,093.6, as compared to $11,082.8 of non-Medicare patients.

SES with Admission Pattern, Severity and Mortality of Pneumonia: An Overall Picture

For all pneumonia patients, as shown in Table 2, associations of socioeconomic status with the admission pattern, severity of illness, and mortality are different. All of the SES measures demonstrated significant associations with both ER admissions and the existence of comorbidity. Among the three measures of race, with non-Hispanic white patients as the reference group, African-American patients showed adjusted ER-admission odds ratios of 1.68 and the existence of comorbidity odds ratio of 1.12, which is consistent with the positive relationship between comorbidity and ER-admission (Boersma, 1999). Relative to non-Hispanic white, other non-African-American patients had ER-admission odds ratio of 1.56, but the existence of comorbidity odds ratio of 0.78. For the insurance status measures, using Medicare patients as the reference group, Medicaid patients and uninsured patients had ER-admission odds ratios of 1.57 and 1.85, but had the existence of comorbidity odds ratios of 0.68 and 0.51, respectively. In other words, as compared to Medicare patients, both Medicaid and the uninsured patients were less likely to have comorbidty, but they were more likely to be admitted through ER, consistent with a lack of accessible continuous care, but less clinically rational. Results of patients covered by private insurance were consistent. As compared to Medicare patients, patients covered by private insurance demonstrated less likelihood to have comorbidity (odds ratio = 0.50) and also less likelihood to experience ER-admission (odds ratio = 0.96). In addition, among the five levels of median income by zip code, the odds ratio of the ER-admission was 1.04 for any of two neighboring levels, from the lower income level to the next higher one.

Results about mortality were relatively less significant. Using non-Hispanic white patients as the reference group, African-American patients showed a higher adjusted mortality odds ratio of 1.13, and the other groups did not vary significantly. Again, as Medicare patients were the reference group, Medicaid patients showed a higher mortality odds ratio of 1.26, while both uninsured

Table 2. ER Admission, Comorbidity, Mortality Risk by SES and Demorgraphic Factors†

	ER Admission ($n = 150{,}071$)			Existence of Comorbidity $n = 153{,}269$			Discharge Died ($n = 153{,}147$)		
	Parameter Estimate	Odds Ratio	95% C.I. for Odds Ratio	Parameter Estimate	Odds Ratio	95% C.I. for Odds Ratio	Parameter Estimate	Odds Ratio	95% C.I. for Odds Ratio
African American	0.527***	1.68	1.61–1.75	0.117***	1.12	1.07–1.18	0.122***	1.13	1.06–1.21
Others	0.434***	1.56	1.49–1.64	–0.254***	0.78	0.73 –0.82	–0.031	1.03	0.95–1.12
Medicaid	0.435***	1.57	1.50–1.66	–0.393***	0.68	0.64–0.72	0.233***	1.26	1.15–1.38
Private insurance	–0.089***	0.96	0.93–1.00	–0.688***	0.50	0.48–0.53	–0.052	0.95	0.89–1.02
Uninsured	0.580***	1.85	1.74–1.98	–0.679***	0.51	0.47–0.54	0.108	1.11	0.98–1.27
Median income by zip code	0.039***	1.04	1.03–1.06	–0.017*	0.98	0.97–1.00	0.007	1.01	0.99–1.03
Age	0.010***	1.01	1.01–1.01	0.353***	1.04	1.04–1.04	0.033***	1.03	1.03–1.04
Female	–0.109***	0.90	0.88–0.92	0.031*	1.03	1.00–1.06	–0.205***	0.82	0.78–0.85

* $p < 0.10$, *** $p < 0.01$

C.I. = Confidence Interval.

† In Tables 2–6, only the results of SES and demographic variables are presented. Full tables containing the results of risk-adjustment variables and the geographic control variables can be obtained from the first author.

patients and patients covered by private insurance did not yield a significant difference in mortality. These results indicate that, after controlling for age, sex, comorbidity, and the geographic variations, African-American pneumonia patients encountered 1.13 times risk of death in hospital as that of non-Hispanic whites, and Medicaid discharges encountered 1.26 times risk of death in hospital as that of Medicare discharges. In addition, the level of the median income by zip code did not seem to be significantly related to mortality.

Pneumonia Mortality: Medicare – A Safety Net

Among Medicare patients, socioeconomic status and ER-admission was significantly related, while the existence of comorbidity was moderately related, and mortality, not significantly related. As shown in Table 3, with non-Hispanic white patients being a reference, both African-American patients and patients of other races had the ER-admission odds ratios of 1.70 and 1.35, respectively, but only the patients of other races showed a significant odds ratio (0.64) for the existence of comorbidity. Both African-American and other racial patients were more likely to be admitted from the emergency room than were non-Hispanic white patients while the other racial patients were less likely to have comorbidity conditions. The median income by zip code was also significantly associated with ER-admission (odds ratio of 1.06) but not significantly associated with the existence of comorbidity. Furthermore, none of the SES measures, both race and the community socioecological condition, showed a significant association with the in-hospital mortality, indicating that the mortality risk of Medicare patients with pneumonia is not different across different levels of socioeconomic status.

Pneumonia Mortality: Non-Medicare and Multiple SES Factors – A Gordian Knot

Without the "universal coverage" of Medicare discharges, non-Medicare discharges' SES measures, with an exception of median income by zip code, were more significantly related to their ER-admission, comorbidity, and especially mortality. Median income by zip code was not associated with ER-admission, comorbidity existence, and mortality. Compared with non-Hispanic white patients, African-American patients showed odds ratios of 1.69, 1.15, and 1.26 for ER-admission, the existence of comorbidity, and hospital mortality, respectively. On the other hand, other racial patients had odds ratios of 1.40, 0.92, and non-significant 0.99 for the same three measures when they were compared against non-Hispanic white patients (Table 4). Among the three

Table 3. ER Admission, Comorbidity, Mortality Risk by SES and Demorgraphic Factors: Medicare Patients with Pneumonia ($n = 90{,}996$)

	ER Admission			Existence of Comorbidity			Discharge Died		
	Parameter Estimate	Odds Ratio	95% C.I. for Odds Ratio	Parameter Estimate	Odds Ratio	95% C.I. for Odds Ratio	Parameter Estimate	Odds Ratio	95% C.I. for Odds Ratio
African American	0.531***	1.70	1.59–1.82	0.099	1.10	0.97–1.26	0.068	1.07	0.99–1.16
Others	0.298***	1.35	1.27–1.43	–0.445***	0.64	0.57–0.71	0.034	1.04	0.95–1.13
Median income by zip code	0.062***	1.06	1.05–1.08	–0.005	0.99	0.96–1.03	0.009	1.01	0.99–1.03
Age	0.004***	1.00	1.00–1.00	0.016***	1.02	1.01–1.02	0.026***	1.03	1.02–1.03
Female	–0.090***	0.91	0.89–0.94	–0.067**	0.94	0.88–0.99	–0.176***	0.84	0.80–0.87

** $p < 0.05$, *** $p < 0.01$
C.I. = Confidence Interval

Table 4. ER Admission, Comorbidity, Mortality Risk by SES and Demorgraphic Factors: Non-Medicare Patients with Pneumonia

	ER Admission			Existence of Comorbidity			Discharge Died		
	Parameter Estimate	Odds Ratio	95% C.I. for Odds Ratio	Parameter Estimate	Odds Ratio	95% C.I. for Odds Ratio	Parameter Estimate	Odds Ratio	95% C.I. for Odds Ratio
Double-Jeopardy variables not included ($n=62,151$)									
African Americn	0.523***	1.69	1.60–1.78	0.139***	1.15	1.09–1.21	0.231***	1.26	1.11–1.43
Other races	0.337***	1.40	1.32–1.48	–0.086***	0.92	0.87–0.97	–0.013	0.99	0.86–1.13
Medicaid	0.481***	1.62	1.55–1.69	0.323***	1.38	1.32–1.45	0.314***	1.37	1.23–1.53
Uninsured	0.628***	1.87	1.76–1.99	0.004	1.00	0.94–1.07	0.191***	1.21	1.05–1.40
Median income by zip code	–0.001	1.00	0.98–1.02	–0.009	0.99	0.97–1.01	–0.001	1.00	0.95–1.05
Age	0.011***	1.01	1.01–1.01	0.038***	1.04	1.04–1.04	0.036***	1.04	1.04–1.04
Female	–0.124***	0.88	0.85–0.92	0.085***	1.09	1.05–1.13	–0.309***	0.73	0.67–0.80
Double-Jeopardy variables included ($n=62,151$)									
Vulnerable race	0.470***	1.60	1.47–1.74	0.309***	1.36	1.25–1.48	0.135	1.15	0.94–1.39
Inferior insurance status	0.580***	1.79	1.71–1.87	0.263***	1.30	1.24–1.36	0.246***	1.28	1.15–1.42
Double-jeopardy interaction	–0.064	0.94	0.85–1.04	–0.196***	0.82	0.74–0.91	0.156	1.17	0.93–1.48
Median income by zip code	–0.010	0.99	0.97–1.01	–0.010	0.99	0.97–1.01	–0.003	1.00	0.95–1.05
Age	0.011***	1.01	1.01–1.01	0.038***	1.04	1.04–1.04	0.036***	1.04	1.04–1.04
Female	–0.127***	0.88	0.85–0.91	0.095***	1.10	1.06–1.14	–0.304***	1.04	0.68–0.81
Comparison of Two Extreme Groups ($n=34,182$)									
Extreme unfavorable	1.095***	2.99	2.79–3.20	0.366***	1.44	1.35–1.54	0.656***	1.93	1.63–2.28
Median income by zip code	0.015	1.02	0.99–1.04	–0.016	0.98	0.96–1.01	0.020	1.02	0.96–1.09
Age	0.013***	1.01	1.01–1.01	0.038***	1.04	1.04–1.04	0.038***	1.04	1.04–1.04
Female	–0.125***	0.88	0.84–0.93	0.091***	1.10	1.04–1.15	–0.288***	0.75	0.67–0.84

*** $p<0.01$.
C.I. = Confidence Interval.

insurance statuses, in comparison with patients covered by private insurance, Medicaid patients showed ER-admission odds ratio of 1.62, the existence of comorbidity odds ratio of 1.38, and mortality odds ratio of 1.37; and uninsured patients showed the odds ratios for the same parameters of 1.87, 1.00, and 1.21, respectively. Based on these results of mortality, African-American race confers more vulnerability than being the non-Hispanic white. Similarly, Medicaid and uninsured are, in fact, inferior insurance statuses compared to private insurance. Results of the three double jeopardy variables, vulnerable race (African-American), inferior insurance status (Medicaid or uninsured), and the interaction of the two are next examined.

Only inferior insurance status shows a significant positive association with ER-admission (odds ratio of 1.79), the existence of comorbidity (odds ratio of 1.30), and mortality (odds ratio of 1.28). Race vulnerability did have significant associations with both ER-admission (odds ratio of 1.60) and the existence of comorbidity (odds ratio of 1.30), but failed to show any significant association with mortality. Further, no significant association between the double-jeopardy interaction and mortality was detected, nor was the association between double jeopardy interaction and ER-admission significant. Nevertheless, a negative association exists between the double-jeopardy interaction and the existence of comorbidity.

Table 4 also presents results of a direct comparison between the two "extreme groups." Comparison of pneumonia discharges belonging to the extremely unfavorable group and the extremely favorable group, respectively, revealed a larger variation in mortality odds, as well as in ER-admission and the existence of comorbidty. The extremely unfavorable patients (list characteristics) had a risk-adjusted mortality odds ratio of 1.93 as opposed to those who were characterized as extremely favorable (list characteristics). Odds ratios of the extremely unfavorable patients over the extremely favorable patients for ER-admission and existence of comorbidities were 2.99 and 1.44, respectively.

Discharges in the extremely unfavorable group encountered a longer length-of-stay (6.4 days) than then the extremely favorable group (5.8 days). Discharges in the extremely unfavorable group incurred higher average total charges per discharge ($11,235.3) than those in the extremely favorable group ($10,440.3) (Table 5).

DISCUSSION

Among the three measures of socioeconomic status, race, insurance, and median income by zip code, the former two are more dominant in relation to

Table 5. Variation in Hospitalization Characteristics between the Extremely Favorable Patients and the Extremely Unfavorable Patients

	Extremely Unfavorable Group		Extremely Favorable Group		Difference Between Two Groups†
	n	Mean (St.d.)	*n*	Mean (St.d.)	
Length of stay	11,837	6.4 (6.1)	27,588	5.8 (5.4)	0.6***
Total charges ($)	11,767	11,235.3 (17,663.7)	26,629	10,440.3 (17,009.9)	795.0***

*** $p < 0.01$.
St.d. = Standard deviation.
† The natural logarithm transformation for all variables was performed before the statistical test.

in-hospital mortality for all pneumonia inpatients analyzed together. Specifically, African-American patients and Medicaid patients are more vulnerable in achieving the same outcome as other patients. Compared to non-Hispanic white patients, African-American pneumonia patients are disadvantaged. They are more likely to have comorbidity, to be hospitalized through ER-admission, and to die during hospitalization. Patients of other races are less likely to have comorbidities but are more likely to experience ER-admission, and are indifferent from non-Hispanic white patients in mortality risk.

Medicaid patients are more disadvantaged in terms of adjusted hospital mortality odds, as compared with patients covered by Medicare. There is no difference in mortality odds ratios among patients having Medicare or private insurance or who are uninsured. Looking at the admission patterns and comorbidity of patients with different insurance statuses, it is apparent that, as compared to Medicare patients, Medicaid patients are less likely to have comorbidity but more likely to have ER-admission and have higher probabilities to experience hospital death. The low likelihood of having comorbidities may result from a much younger average age of Medicaid patients than that of Medicare patients. The higher probability for ER-admission and mortality risk may indicate that the delayed care-seeking behavior might result in the worsening post-admission outcome. When patients under private insurance and Medicare patients are compared, the former are less likely to be admitted through ER and have comorbidities, but do not achieve a lower adjusted-mortality risk. This finding may be explained by the observation that younger patients, covered by private insurance, have less comorbidity and, lower chances of ER presentation. The reason that Medicare patients and patients covered by private insurance are indifferent in pneumonia mortality risk is that the mortality is clinically risk-adjusted. Without adjusting comorbidities and other risk factors, Medicare patients did have a higher mortality risk (results not presented here). Comparing uninsured patients against Medicare patients reveals an interesting phenomenon. Despite being less likely to have comorbidities, the uninsured experience more ER-admission and then achieve a similar outcome for mortality risk. Further investigation is needed to examine whether the uninsured pneumonia patients with higher ER-admission rates offsets some of their disadvantages in achieving outcome similar to those of Medicare patients.

The median income by zip code is positively associated with more ER-admission but is not statistically significantly related to hospital mortality. One possible explanation is that, after controlling for race, other insurance status, age, gender, clinical risk factors, and geographic regions, relatively more Medicare patients live in areas that have relatively higher median income by zip code and in which relatively more patients covered by private insurance live. Compared to patients covered by private insurance, Medicare patients are more likely to have comorbidities and, therefore, are more likely to require ER services.

The Medicare program receives a favorable support by findings of this study. This study reveals that Medicare program provides a good "safety net" for its beneficiaries who suffer from pneumonia and being hospitalized, against hospital mortality risk. Although variations exist in comorbidity conditions and in ER-admission, no significant difference in mortality risk is detected across different racial/ethnic groups, patients with different insurance status, and patients living in areas having different median income by zip code. Medicare program achieves a satisfactorily "homogeneous" outcome for all of its hospitalized beneficiaries with pneumonia regardless their socioeconomic status.

The relationship between socioeconomic status and outcome among non-Medicare patients with pneumonia exists but is more complicated. As compared to non-Hispanic white patients, African-American patients are more likely to have comorbidities, ER-admission, and hospital death. Patients of other races are less likely to have comorbidities but more likely to experience ER-admission and have similar hospital mortality risk. Among the three racial groups, non-Hispanic whites, African-Americans, and others, African-Americans with pneumonia are the only group that is disadvantaged with respect to the outcomes explored. One reason for experiencing unfavorable outcome is the higher probability of suffering comorbidities (not related to pneumonia), which may indicate worse general health status associated with certain socioeconomic factors such as healthy behavior, life style, and social barriers precluding adequate and necessary care. The observation that patients of other races, despite having less comorbidity, experience more ER-admissions, deserves further study. One explanation for this result is that this group lacks a usual care source (Agency for Health Care Policy and Research, 1998). Once being sick (e.g. pneumonia), they will go directly to the ER without physician consultation or referral (Blustein, Hanson & Shea, 1998).

The higher mortality risk demonstrated for Medicaid patients and uninsured patients over that of patients covered by private insurance may indicate different socioeconomic disadvantages. For Medicaid patients, the higher mortality risk may largely result from both higher occurrence of non-pneumonia related comorbidities and higher ER-admission rates that often reflect more severe cases (Elixhauser et al., 1998). For uninsured patients, the worse mortality risk may result primarily from financial barriers to have usual care sources such as physician offices, outpatient settings, and non-ER admission. Having no health insurance coverage increases the use of a non-ambulatory-care site as a place for usual care (O'Toole et al., 1999). Due to financial constraints, uninsured patients may wait until their pneumonia is quite severe and then require emergency care (Fraser, 1997). The results of this study display the highest ER-admission rates among all types of insurance status in the uninsured.

It is worth noted that, using private insurance as a reference, Medicaid had a higher mortality risk than uninsured patients (odds ratios: 1.37 vs. 1.21 in this study). Financially, Medicaid patients hold advantages over uninsured patients because the uninsured status may implicate the lack of access to care imposed by financial barriers (Gold, 1998; Schoen & DesRoches, 2000). As Medicaid patients are qualified by income, uninsured patients are excluded from Medicaid but often cannot afford health insurance. Uninsured patients are likely unprepared for the out of pocket expense of pneumonia for which total charges per episode is approximately $11,000 in this study. As a result of delaying care, uninsured patients may require more frequent ER-admissions (odds ratios: 1.87 vs. 1.62). In contrast, uninsured patients have advantages over Medicaid patients in having less comorbidities (odds ratios: 1.00 vs. 1.38) and probably better health status. Although both severity of illness and financial barriers contribute to higher mortality risk, it appears that the former does so more than the latter.

Median income by zip code areas, as the measure of socioecological conditions of a community, is not associated with pneumonia mortality, ER-admission, or comorbidity conditions, for non-Medicare patients. This community-level measure may not be sensitive enough to identify the individual socioeconomic differences. Also, since there was a degree of correlation between median income by zip code and the two individual-level socioeconomic measures, race and insurance status, the relationship between median income by zip code and outcome may be overshadowed by those between race and outcome and between insurance status and outcome.

The double-jeopardy hypothesis that both vulnerable race and inferior insurance confers an increased risk for pneumonia mortality did not receive statistically significant support from this study. Despite the fact that patients with the inferior insurance status (Medicaid or uninsured) have higher mortality risk than do patients covered by private insurance, patients of vulnerable race (African-American) do not have mortality risk than do non-Hispanic white patients. Moreover, the interaction term of the two double-jeopardy measures is also not statistically significant. It is not observed that the increase in mortality risk as insurance status changes from private insurance to the inferior insurance status is greater for African-American patients than for non-Hispanic white patients. If the hypothesis is true, one reason for not detecting any significant double jeopardy phenomenon related to pneumonia mortality may be an underlying relationship between inferior insurance status and vulnerable race. This relationship may not be simply linear but could be multi-linear (showing linear correlation after controlling for other factors) through which the inferior insurance status significantly aggravates survival disadvantage. It implies again that race/ethnicity is a very complex measure which represents multiple underlying socioeconomic elements, as well as biological and cultural differences (Belgrave, Wykle & Choi, 1993; HCFA, 1999; Mueller, Patil & Boilesen, 1998; Schulman, 1995). Future studies evaluating the relationship between insurance and race can elucidate more specific socioeconomic factors reflected by race. Successes of this effort will enable researchers to select more specific and sensitive socioeconomic measures to rectify the double jeopardy hypothesis.

The absence of the double-jeopardy effects on pneumonia patients in this study does not assuage concerns about the outcome of those patients who have more than one socioeconomic disadvantage. Findings of this study indicate that two socioeconomic risk factors associated with worsen outcomes for pneumonia patients. The extremely unfavorable risks expressed in pneumonia patients, as opposed to the extremely favorable risks, confer much higher mortality risks, higher rates of ER-admission, more comorbidities, longer average length-of-stay (likely a result of multiple comorbidities and more advanced disease), and higher average total charges. From the societal perspective, the current dynamics of treating pneumonia patients with multiple socioeconomic advantages seems less than ideal. Due to socioeconomic barriers, socioeconomically disadvantaged patients arrive in hospital emergency rooms with serious illnesses that could have been treated less expensively and more successfully if they were treated earlier (Fraser, 1997).

Regardless of comorbidities and mortality, socioeconomically vulnerable pneumonia patients tend to seek more ER-admissions. Minority patients are likely to have more ER-admissions than are non-Hispanic white patients, and both Medicaid patients and uninsured patients are likely to have more ER-admissions than are those patients under private insurance. Several factors may explain this observation. First, minority patients and patients with relatively inferior insurance status may be more likely to have financial barriers and other socioeconomic barriers, such as transportation problems, communication problems, and physical problems, to receive timely and necessary care (Agency for Health Care Policy and Research, 1998). Circumstance may result in waiting until the disease symptoms get quite severe which then requires seeking care in the emergency room (Fraser, 1997; White-Means, 1995). The delay in obtaining care results in more severe clinical conditions that require hospitalization. Second, socioeconomically vulnerable patients (minorities or uninsured) more often lack an usual source of health care; therefore, hospital services, especially the emergency care, become a more frequently used source of care for all healthcare needs (Agency for Health Care Policy and Research, 1998; O'Toole et al., 1999; White-Means, 1995). Recent data reveal that more Hispanic and uninsured patients do not have usual source of continuity care resulting in use of more ER services (Agency for Health Care Policy and Research, 1998). Findings of this study show consistent results. Although both non-African-American minorities and uninsured do not suffer more comorbidity conditions than do non-Hispanic whites and patients covered by private insurance, respectively, they are more likely to be admitted from the ER. Hospital services, especially emergency care, currently offer an important source of care to those socioeconomically vulnerable pneumonia patients (Gaskin, 1998). Although it is assumed that most of these hospital services rely on public support or charity, in an era of greater financial accountability with a concomitant reduction of cost-shifting historically utilized to support indigent care, further studies are needed to examine the association between characteristics of hospital services and socioeconomically disadvantaged patients. In any case, those hospital services deserve financial supports to meet some of health care needs of socioeconomically disadvantaged populations.

Results of this study about Medicaid patients and pneumonia outcome do not strengthen findings from other studies. Previous evidence has shown that Medicaid patients have better sources of usual care and better access to care than do uninsured patients (Agency for Health Care Policy and Research, 1998; Berk & Schur, 1998; Schoen & DesRoches, 2000). The results of this study, however, do not demonstrate better outcomes of Medicaid patients than that of the uninsured patients. Compared to patients under private insurance, Medicaid patients even had a slightly higher mortality odds ratio than did the uninsured patients. This phenomenon partly derives from the fact that Medicaid patients have higher likelihood of comorbidities. It may imply underlying socio-economic risk factors beyond the conventional access boundary often measured by insurance coverage. Poverty of Medicaid covered patients may be compounded by other factors such as education, health behavior, living condition, environmental exposure, and even early life conditions (Lantz et al., 1998; Williams & Collins, 1995), all factors that can result in worse health outcomes.

Several limitations exist in the methodology employed in this study. First, the UB-92-based NIS data lack richly detailed clinical diagnosis and procedure information. As noted, some clinical risk measures of pneumonia obtained from clinical laboratory and physical examination findings are not available in the NIS data. These clinical data have shown significant effect on severity of pneumonia and mortality (Agency for Health Care Policy and Research, 1997; Meehan et al., 1997). Also, insufficient clinical procedure information available for pneumonia in the NIS dataset does not allow the comparison of process of care received by patients across different socioeconomic classifications, although the process of care differentiates the outcome of pneumonia (Fine et al., 1999; Meehan et al., 1997). Second, in addition to hospital mortality, other outcome measures such as post-discharge mortality or complications might allow comparison of the consistency of the associations between socio-economic status and outcome. However, post-discharge mortality information is not identified in the NIS data. This study did intend to use complication as another outcome measure that is operationalized by the complication list of the ICD-9-CM codes developed by DesHarnais (1990). However, since the complication list only contains those conditions that are truly certain to be complications and does not include those that can be either complications or comorbidities, it under-identifies complication, resulting fairly low complication rate (around 1.3%) which rendered the multivariate likelihood estimation non-convergent. Finally, exclusion of discharges in the six states due to lack of necessary information may reduce the generalizability of the results of this study.

Policy Implications

Substantial variation across socioeconomic status in the patterns of seeking care and the outcome for the treatment of pneumonia would seem to be central to the national discussion regarding access to and the quality of healthcare. One policy strategy might be to continue improving the two public health insurance programs, Medicare and Medicaid. Due to the economic concern about the future viability of Medicare, approaches are needed to maintain and prolong the program. This research demonstrates that Medicare provides a safety-net, for its beneficiaries with pneumonia, which reduces negative variation in the outcome across different socioeconomic strata. Medicaid programs or other governmental programs may be augmented by work addressing the socio-economic factors that lead to high comorbidities for all Medicaid patients and African-American patients, in particular. Fewer comorbidities will result in better outcomes in pneumonia and other diseases.

Second, negative conclusions regarding the efficiency and effectiveness of hospital admissions through emergency room should be made with caution. Due to lack of continuity of care for minority and uninsured populations, hospitals, especially the emergency room, provide an episodic source of care for those population groups for acute needs (Agency for Health Care Policy and Research, 1998). Until better approaches are formed to improve continuity care for socioeconomically vulnerable populations, some "over-utilization" of the ER-admission constitutes a realistic way to provide care and produce the best outcomes available. Continuous public support of this "non-efficient" but necessary source of care is required in absence of solutions addressing the underlying issues

Similarly, treatment of pneumonia patients with low socioeconomic characteristics may be improved from a cost-effectiveness perspective. The results of this study show that socioeconomically vulnerable patients underwent more comorbidities, more ER-admissions, greater length of hospital stays, higher charges, and worse clinical outcomes. The same findings are observed in other diseases (Shen, Wan & Perlin, 2001). Again, the cost-effectiveness justification should consider both access and equity of care among persons across different socioeconomic statuses.

In conclusion, inequity in health care and outcome remains an outstanding issue in our society. Understanding of relationships among socioeconomic status, insurance status, and race classification can assistant in seeking more effective policy strategies as well as in developing more effective programs to improve access to care and achieve better clinical and economic outcomes. Greater and enduring efforts, from both public and private sectors, are required to explore better approaches to improve the access and care for socio-economically disadvantaged populations in our society; in particular, those who tolerate multiple socioeconomic risk factors associated with unfavorable outcomes.

REFERENCES

Agency for Health Care Policy and Research (AHCPR) (now changed to Agency for Health Research and Quality, AHRQ) (1998). *Access to Health Care – Sources and Barriers, 1996: Research Findings No. 3.* AHCPR Pub. No. 98–0001.

Agency for Health Care Policy and Research (AHCPR) (now changed to Agency for Health Quality and Research, AHQR) (1997). *New Prediction Model Proves Promises.* AHCPR Pub. No. 97-R031.

Altman, S. H., Reinhardt, U. E., & Shields, A. E. (Eds) (1998). *The Future U.S. Healthcare System: Who Will Care for the Poor and the Uninsured?* Chicago, IL: Health Administration Press.

American Thoracic Society (1993). Guidelines for the Initial Management of Adults with Community-Acquired Pneumonia: Diagnosis, Assessment of Severity, and Initial Anti-microbial Therapy. *American Reviews of Respiratory Diseases, 148,* 1418–1426.

Ayanian, J. Z., Weissman, J. S., Chasan-Taber, S., & Epstein, A. M. (1999). Quality of Care by Race and Gender for Congestive Heart Failure and Pneumonia. *Medical Care, 37*(12), 1260–1269.

Baker, D. W., Stevens, C. D., & Brook, R. H. (1996). Determinants of Emergency Department Use: Are Race and Ethnicity Important? *Annals of Emergency Medicine, 28,* 677–682.

Belgrave, L. L., Wykle, M. L., & Choi, J. M. (1993). Health, Double Jeopardy, and Culture: The Use of Institutionalization by African-Americans. *The Gerontologist, 33*(3), 379–385.

Berk, M. L., & Schur, C. (1998). Access to Care: How Much Difference Does Medicaid Make? *Health Affairs, 17*(3), 169–180.

Bloomberg, L., Meyers, J., & Braverman, M. T. (1994). The Importance of Social Interaction: A New Perspective on Social Epidemiology, Social Risk Factors, & Health. *Health Education Quarterly, 21*(4), 447–463.

Blustein, J., Hanson, K., & Shea, S. (1998). Preventable Hospitalizations and Socioeconomic Status. *Health Affairs, 17*(2), 177–189.

Boersma, W. G. (1999). Assessment of Severity of Community-Acquired Pneumonia. *Seminars in Respiratory Infections, 14*(2), 103–114.

Charlson, M.E, Pompei, P., Ales, K. L., & MacKenziek C. R. (1987). A New Method of Classifying Prognostic Comorbidity in Longitudinal Studies: Development and Validation. *Journal of Chronicle Diseases, 40*(5), 373–383.

Cludio, L., Tulton, L., Doucette, J., & Landrigan, P.,J. (1999). Socioeconomic Factors and Asthma Hospitalization Rates in New York City. *Journal of Asthma, 36*(4), 343–350.

DesHarnais, S. (1990). Current Uses of Large Data Sets to Assess the Quality of Providers. Construction of Risk-Adjusted Indexes of Hospital Performance. *International Journal of Technology Assessment in Health Care, 6,* 229–238.

Deyo, R. A., Cherkin, D. C., & Clol, M. A. (1992). Adapting a Clinical Comorbidity Index for Use with ICD–9-CM Administrative Databases. *Journal of Clinical Epidemiology, 45*(6), 613–619.

Elixhauser, A, Steiner, C, Harris, D. R., & Coffey, R. M. (1998). Comorbidity Measures for Use with Administrative Data. *Medical Care*, *36*(1), 8–27.

Ferraro, K. F. (1989). Reexaming the Double Jeopardy to Health Thesis. *Journal of Gerontology*, *44*(1), 514–516.

Fine, M. J., Smith, M. A., Carson, C. A., Mutha, S. S., Sankey, S. S., Weissfeld, L. A., & Kapoor, W. N. (1996). Prognosis and Outcomes of Patients with Community-Acquired pneumonia. A Meta-Analysis. *Journal of the American Medical Association*, *275*(2), 134–141.

Fine, M. J., Stone, R. A., Singer, D. E., Coley, C. M., Marrie, T. J., Lave, J. R., Hough, L. J., Obrosky, D. S., Schulz, R., Ricci, E. M., Rogers, J. C., & Kapoor, W. N. (1999). Process and Outcomes of Care for Patients with Community-Acquired Pneumonia. *Archives of Internal Medicine*, *159*, 970–980.

Fine, M. J., Auble, T. E., Yealy, D. M., Hanusa, B. H., Weissfeld, L. A., Singer, D. E., Coley, C. M., Marrie, T. J., & Kapoor, W. N. (1997). A Prediction Rule to Identify Low-Risk Patients with Community-Acquired Pneumonia. *The New England Journal of Medicine*, *336*(4), 243–250.

Fraser, I. (1997). Access to Health Care. In: T. J. Litman & L. S. Robins (Eds), *Health Politics and Policy, Third Edition* (pp. 288–305). Albany, NY: Delmar Publishers.

Garibaldi, R. A. (1985). Epidemiology of Community-Acquired Respiratory Tract Infections in Adults: Incidence, Etiology, & Impact. *American Journal of Medicine*, *78*, 32S–37S.

Gaskin, D. J. (1998). The Hospital Safety Net: A Study of Inpatient Care for Non-Elderly Vulnerable Populations. In: M. Lillie-Blanton, R. M. Martinez, B. Lyons & D. Rowland (Eds), *Access to Health Care: Promises and Prospects for Low-Income Americans* (pp. 123–138). Washington, D.C.: The Henry J. Kaiser Family Foundation.

Gold, M. (1998). Beyond Coverage and Supply: Measuring Access to Healthcare in Today's Market. *Health Services Research*, *33*(3, Part II), 625–652.

Haywood, L. J., Ell, K., deGuman, M. et al. (1993). Chest Pain Admissions: Characteristics of Black, Latino, & White Patients in Low- and Mid-Socioeconomic Strata. *Journal of National Medical Association*, *85*(10), 749–757.

Health Care Financing Administration (HCFA) (1999). http://www.hcfa.gov/medicaid/2082-8.htm.

La Force, F. M. (1985). Community-Acquired Lower Respiratory Tract Infections: Prevention and Cost-Control Strategies. *American Journal of Medicine*, *78*, 52–57.

Lantz, P. M., House, J. S., Lepkowski, J. M., Williams, D. R., Mero, R. P., & Chen, J. M. (1998). Socioeconomic Factors, Health Behaviors, & Mortality: Results from a Nationally Representative Prospective Study of US Adults. *Journal of the American Medical Association*, *279*(21), 1703–1708.

Lillie-Blanton, M., Parsons, P. E., Gayle, H., & Dievler, A. (1996). Racial Differences in Health: Not Just Black and White, but Shades of Grey. *Annual Review of Public Health*, *17*, 411–448.

Lillie-Blanton, M. (1999). A Review of the Nation's Progress and Challenges in Assuring Access to Health Care for Low-Income Americans. In: M. Lillie-Blanton, R. M. Martinez, B. Lyons & D. Rowland (Eds), *Access to Health Care: Promises and Prospects for Low-Income Americans* (pp. 19–37). Washington, D.C.: The Henry J. Kaiser Family Foundation.

Meehan, T. P., Fine, M. J., Krumholz, H. M., Scinto, J. D., Galusha, D. H., Mockalis, J. T., Weber, G. F., Petrillo, M. K., Houck, P. M., & Fine, J. M. (1997). Quality of Care, Process, and Outcomes in Elderly Patients with Pneumonia. *The Journal of American Medical Association*, *278*(23), 2080–2084.

Moor, K. (1997). Socioeconomic Status as Quarantine: Unfortunate Consequences of the Inability to Afford Health Care. *Outcomes Management of Nursing Practice, 1*(1), 31–38.

Mueller, K. J., Patil, K., & Boilesen, E. (1998). The Role of Uninsured and Race in Healthcare Utilization by Rural Minorities. *Health Services Research, 33*(3), 597–610.

Nickens, H. W. (1995). The Role of Race/Ethnicity and Social Class in Minority Health Status. *Health Services Research, 30*(1, Part II), 151–162.

Niederman, M. S., McCombs, J. S., Unger, A. N., Kumar, A., Popovian, R. (1998). The Costs of Treating Community-Acquired Pneumonia. *Clinical Therapy, 20*(4), 820–837.

O'Toole, T. P., Gibbon, J. L., Hanusa, B. H., & Fine, M. J. (1999). Preferences for Sites of Care among Urban Homeless and Housed Poor Adults. *Journal of General Internal Medicine, 14*(10), 599–605.

Papps, G., Hadden, W. C., Kozak, L. J., & Fisher, G. F. (1997). Potentially Avoidable Hospitalizations: Inequalities in Rates between US Socioeconomic Groups. *American Journal of Public Health, 87*(5), 811–816.

Rich, D. W., Rich, T., & Mullins, L. C. (Eds) (1995). *Old and Homeless – Double-Jeopardy: An Overview of Current Practice and Policies.* Westport, CN: Auburn House.

Salmon, M. A. P. (1994). *Double Jeopardy: Resources and Minority Elders.* New York: Garland.

SAS Institute Inc. (1990). *SAS/STAT User's Guide, Volumes 1 and 2.* Cary, NC: SAS Institute Inc..

SAS Institute Inc. (1997). *SAS/STAT Software: Changes and Enhancements through Release 6.12.* Cary, NC: SAS Institute Inc..

Schoen, C., & DesRoches, C. (2000). Uninsured and Unstably Insured: The Importance of Continuous Insurance Coverage. *Health Services Research, 35*(1, Part II), 187–206.

Schulman, K. A., Rubenstein, L. E., Chesley, F. D., & Eisenburg, J. M. (1995). The Roles of Race and Socioeconomic Factors in Health Services Research. *Health Services Research, 30*(1, Part II), 179–195.

Shen, J. J., Wan, T. T. H., & Perlin, J. B. (2001). An Exploration of the Complex Relationship of Socioecological Factors in the Treatment and Outcomes of Acute Myocardial Infarction in Disadvantaged Populations. *Health Services Research*, in press.

Smith, G. D., Wentworth, D., Neaton, J. D. et al. (1996). Socioeconomic Differentials in Mortality Risk Among Men Screened for the Multiple Risk Factor Intervention Trial: II. Black Men. *American Journal of Public Health, 86*(4), 497–504.

Smith, M. and Chang Pancho. (1995). Editorial Preface. *Health Services Research, 30*(1, Part II), 145–149.

White-Means, S. I. (1995). Conceptualizing Race in Economic Models of Medical Utilization: A Case Study of Community-Based Elders and the Emergency Room. *Health Services Research, 30*(1, Part II), 207–223.

Williams, D. R., & Collins, C. (1995). U.S. Socioeconomic and Racial Differences in Health: Patterns and Explanations. *Annual Review of Sociology, 21*, 349–386.

Young, G. J., & Cohen, B. B. (1991). Inequities in Hospital Care, the Massachusetts Experience. *Inquiry, 28*, 255–262.

SOCIOECONOMIC STATUS AND HEALTH BEHAVIORS AMONG CALIFORNIANS

Joseph G. Grzywacz

ABSTRACT

Despite evidence indicating that socioeconomic status is multidimensional, health behavior research frequently "controls for" education or income as monolithic indicators of SES. Education and income (as well as other SES indicators) tap different aspects of SES, are only moderately correlated, and can each independently influence behavior. The specific aims of this study were to: (1) describe and disaggregate the association between different indicators of socioeconomic status –educational attainment, employment status (as a proxy for occupational status), individual income, and household income—and multiple health behaviors; (2) examine the pattern of association between SES and health behaviors by gender; and (3) explore different hypotheses linking SES and health behaviors. In this study, like previous work, educational attainment had the most consistent relationship with different health behaviors, particularly among men. However an interesting pattern of associations between different dimensions of SES and different health behaviors emerged in multivariate analyses, as did several gender differences. Although fewer psychological resources (i.e. higher levels of depression and lower feelings of control) explained the SES disadvantage in terms of sedentary living, little evidence of mediation was found for other

Health, Illness, and Use of Care: The Impact of Social Factors, Volume 18, pages 121–149.

ISBN: 0-7623-0740-4

SES-health behavior linkages. Results from this study suggest that multiple aspects of socioeconomic disadvantage can undermine various behaviors related to health. Health interventions targeting lower status populations may need to incorporate specific strategies addressing different aspects of social class.

INTRODUCTION

An individual's participation in salutary health practices, like health status, is intimately linked to her/his location in the social hierarchy: lower status individuals usually engage in fewer positive health practices and more health-risk behaviors than higher status individuals (National Center for Health Statistics [NCHS], 1998). Indeed, the lifestyle hypothesis is one of the primary mechanisms that have been offered to explain the socioeconomic status-health relationship (for reviews see Adler et al., 1994; Anderson & Armstead, 1995; Evans, Barer & Marmor, 1994; Feinstein, 1993; Haan, Kaplan & Camacho, 1989; Lantz et al., 1998; Williams, 1990). Unlike the SES-health relationship however, the SES and health behavior relationship remains under-researched (Gochman, 1997; Pol & Thomas, 1997), with some scholars contending that indiscriminately "controlling for" SES is a major problem in public health research (Marmot, Kogevinas & Elston, 1987). The purpose of this study was to systematically examine the association between different dimensions of SES and different health behaviors emphasized in the Healthy People 2000 objectives (Public Health Service [PHS], 1991). Further, four different mechanisms through which SES may influence behaviors related to health were explored.

THEORETICAL AND EMPIRICAL BACKGROUND

Socioeconomic status is a multidimensional latent construct representing an individual's location in the social hierarchy as reflected by her/his ability to create or consume valued goods (Hauser & Warren, 1996). Typically observed indicators of latent constructs are conceptualized as effects of the latent factor (i.e., "effect indicator" constructs; Bollen, 1984), however SES is best described as a "cause indicator" latent construct (Bollen, 1984). Education, earnings, and other frequently used indicators of SES are not necessarily caused by SES, rather individuals' location in status hierarchies emerge from or are caused by different factors (e.g., education and earnings) that facilitate the creation or consumption of valued goods. Unlike "effect indicator" latent constructs that have favorable psychometric properties, observed indicators (e.g., education and earnings) for "cause indicator" constructs (e.g., SES) can be positively correlated, negatively correlated, or have zero correlation but yet be valid and important indicators of the construct (Bollin & Lennox, 1991).

Although recent reviews emphasize the importance of considering the independent effects of multiple aspects of socioeconomic status in health-related research (Krieger, Williams & Moss, 1997; Williams, 1990), most studies of different health behaviors rely primarily on education or income as unitary measures of SES (Sobal & Stunkard, 1989). Because of the latent structure of SES however, different indicators have the potential of exerting independent, unique effects on health-related behaviors (Marks, 1996; Slater, Lorimor & Lairson, 1985; Williams, 1990). Research examining the independent influence of multiple indicators of SES can provide a better, more comprehensive understanding of health-related behaviors by identifying the "net effect" of different aspects of SES on health behaviors (Marks & Shinberg, 1998). The "net effects" of SES can then be used to categorize individuals into comprehensive typologies for considering risk, as well as to elucidate more specific SES-health behavior linkages.

SES and Health Behaviors

Descriptive reports from countries around the world note a similar pattern in the association between unique dimensions/indicators of socioeconomic status and health behaviors. Studies examining differences in health behaviors by occupational class or status find that women and men in lower class, manual occupations are more likely to smoke, to be obese, and to lead a sedentary lifestyle than individuals in higher class, non-manual occupations (Blaxter, 1990; Marmot et al., 1991; Marmot et al., 1997; Pocock, Shaper, Cook, Phillips & Walker, 1987; Wray, Herzog, Willis & Wallace, 1998). Similarly, lower status individuals, in terms of education *or* income, are more likely to smoke, to smoke more cigarettes per day, to be obese, to experience a significant weight gain, and to be less physically active than individuals with a higher level of education or income (Dannenberg, Keller, Wilson & Castelli, 1989; Kahn, Williamson & Stevens, 1991; Krick & Sobal, 1990; Matthews, Kelsey, Meilahn, Kuller & Wing, 1989; Millar & Wigle, 1986; NCHS, 1993; NCHS, 1998; Sobal & Stunkard, 1989). In other words, descriptive studies measuring individual dimensions of SES consistently find that lower status individuals participate in fewer positive health behaviors and more health-risk behaviors.

Notable deviations from the consistent pattern of relationships between socioeconomic status and different health behaviors exist however and require attention. Two studies examining SES differences in the number of kilocalories expended in physical activity found a clear gradient for women, but results were less clear among men (Folsom et al., 1985; Ford et al., 1991). In both studies women with a higher level of education expended more total energy in leisure time, job-related, and household-related physical activities. However, among young men (aged 25–44), the amount of energy expended in leisure time physical activity was highest for those with the lowest education, while the typical gradient emerged for men 45 years of age or older. Also among men, a U-shaped association was found for participation in heavy leisure-time physical activity (Folsom et al., 1985); otherwise, no systematic differences in physical activity by SES was found (Ford et al., 1991).

The dissimilarity between the pattern of results in these two studies of exercise in contrast to those previously cited draw attention to some important features of the socioeconomic status and health behavior relationship. First the notable gender differences in the effects of SES on exercise-related outcomes supports arguments that typical conceptualizations of SES are gendered (McDonough, Williams, House & Duncan, 1999), and raise the possibility that the effects of SES on other health behavior outcomes may also differ by gender. These studies also raise caution flags regarding operational definitions of health behavior outcomes. While measurement of some health behaviors is clear (e.g. in contrast to smoking, smoking at any level undermines health; Doll & Peto, 1976), operationalizing other health behaviors, particularly exercise (Dishman, 1994) is less clear. Using everyday forms of physical activity, also important to adult health (Andersen et al., 1999; Dunn et al., 1999), these studies found a different pattern of effects by SES in contrast to studies using status-based indicators of physical activity (e.g. participation in sporting events or leisure time physical activity). In other words, these results suggest that that some findings in the literature may be an artifact of measure selection.

Another interesting pattern of results emerges from empirical descriptions of the association between socioeconomic status and alcohol-related outcomes. Individuals with a fewer years of schooling report more problem drinking and more symptoms of dependence (Hilton, 1990a; Kohn, Dohrenwend & Mirotznik, 1998; Matthews, Kelsey, Meilahn, Kuller & Wing, 1989; Millar & Wigle, 1986; NCHS, 1998), but individuals with higher incomes and greater occupational status have been found to drink more often, have higher levels of alcohol consumption, and to report more negative social consequences resulting from drinking (Hilton, 1990a; Marmot et al., 1991; Matthews, Kelsey, Meilahn, Kuller & Wing, 1989). Despite the apparent contradiction, emerging consensus is forming suggesting that higher status individuals drink more regularly but they consume modest levels per setting while lower status individuals who drink are more likely to be problem or risky drinkers. Indeed, once analyses are restricted to individuals who drink regularly, the inverse relationship between income and alcohol-related outcomes reverses such that lower income respondents report more problems (Hilton, 1990b).

The association between socioeconomic status and alcohol-related outcomes appears to be conditioned by gender (Hilton, 1990a; Kohn, Dohrenwend & Mirotznik, 1998). Not only do men drink more and report more symptoms of dependence than women (Grant et al., 1994; U.S. Department of Health and Human Services, 1997), but the link between different SES and alcohol-related outcomes has been found to hold only for men (Hilton, 1990a, 1990c; Kohn, Dohrenwend & Mirotznik, 1998). These results suggest yet again that the impact of SES on different health behaviors may differ for women and men.

Multiple Dimensions of SES and Health Behaviors

Associations between socioeconomic status and different health behaviors are usually described using one measure or another of socioeconomic status at a time – usually education or income (e.g. Sobal & Stunkard, 1989). The use of education or income as monolithic indicators of SES is often forced by restrictions of available measures. However, research examining the unique effects of multiple indicators of SES on health behaviors can, among other things, guide the search for more proximate causal linkages underlying the relationship and facilitate more elaborate models of health behavior change (Kessler, 1982; McLeod & Kessler, 1990; Mechanic, 1989). Only a few studies have simultaneously modeled the effects of multiple indicators of SES on health behaviors (Grzywacz & Marks, Forthcoming; Marks & Shinberg, 1997, 1998; Wray et al., 1998) and even fewer studies have examined multiple health behaviors (Marks, 1996).

Studies examining the unique effects of different dimensions of socio-economic status usually find that education is the most important aspect of SES in health-related behaviors, but other dimensions of SES have been found to exert parallel independent, unique effects on different health behaviors (Marks, 1996; Wray et al., 1998). Moreover, models examining the unique effects of different aspects of SES occasionally yield significant, unique effects that are contradictory. For example, a forthcoming study of problem drinking during midlife found that less education was associated with higher odds of problem drinking but lower income was associated with lower odds of problem drinking net of education, and this pattern did not differ by gender (Grzywacz & Marks, 2000). Similarly longitudinal evidence indicated that lower *education* was associated with *more* smoking among women but lower income, net of education, was associated with *less* smoking among women (Marks, 1996).

These results illustrate the complexity of the SES-health behavior relationship and indicate that additional research is necessary to help explain the effects of different aspects of SES on behaviors related to health.

Explanations for the SES-Health Behavior Relationship

Paralleling the larger socioeconomic status-health literature, several underlying causal linkages can be offered to explain the relationship between SES and health behaviors. Physical or social strains/barriers (e.g. neighborhood safety; Centers for Disease Control, 1999a) as well as psychological stress (e.g. work stress; Conway, Vickers, Ward & Rahe, 1981) influences participation in a variety of health behaviors. Therefore differential exposure to *environmental strains or life stresses* may explain differences in health practices between higher and lower status individuals. Since social support is important for adopting, maintaining, or changing health-related behaviors (e.g. Berkman, 1995) differences in the level of *social integration and the quality of social ties* may explain differences in health behaviors. Differences in the relative level of various *psychological resources* (e.g. control over health; Strickland, 1978) may also explain variability in lifestyle practices by SES. Finally, since health care providers are important sources for information regarding health behaviors, access to medical care may provide another potential explanation for differences in behavior related to health.

The goals of this study were to further our understanding of the social etiology of health-related behaviors by: (1) describing and disaggregating the association between different indicators of socioeconomic status – educational attainment, employment status (as a proxy for occupational status), individual income, and household income – and multiple health behaviors; (2) examining the pattern of association between SES and health behaviors by gender; and (3) exploring different hypotheses linking SES and health behaviors.

METHODS

Data

Data from the first wave of the California Work and Health Survey (1998) were used in this study. The CHWS is a telephone based longitudinal survey of 1,771 California adults developed by the faculty and staff of the Work and Health Program at the University of California, San Francisco, with input from researchers and practitioners in the fields of health and economics. The survey was conducted by the Field Institute and funded by a grant from the California

Wellness Foundation under the Work and Health Initiative. The survey is unique in its extensive coverage of employment status and working conditions, as well as different aspects of physical and mental health. Sampling weights allow this sample to match the general California population in terms of household size, age, gender, race/ethnicity, and region.

Using random digit dialing procedures and oversampling strategies for African Americans, Asian/Pacific Islanders, and persons with disability, 8,848 persons were contacted from selected residential telephone numbers. Up to six attempts were made to reach a person at the selected telephone number. Of the persons contacted, approximately 3,106 were eligible for inclusion in the survey yielding a response rate of approximately 57% (i.e. 1,771/3,106). The overall cooperation rate for this sample is somewhat lower than might be expected: this was partially due to the oversampling techniques which yielded a high number of terminated interviews (2,460) since the contact did not meet the eligibility criteria for their particular sample augment (i.e. African American, Asian/Pacific Islander, or disabled). (For additional information on the CWHS see http://medicine.ucsf.edu/programs/cwhs.)

Measures

Health Behaviors Outcomes

Different aspects of smoking, exercise/fitness, and alcohol were considered in this study. Self-reported smoking status (1 = smoker) and average number of cigarettes smoked per day were used to assess smoking-related behavior. Sedentary lifestyle ("During the past month, did you participate in any physical activities or exercises such as running, calisthenics, golf, gardening, or walking for exercise?"; 1 = No), regular participation in aerobic activities (individuals reporting running, jogging, walking or swimming three or more times per week are coded 1, 0 otherwise), and obesity (Body Mass Index > 27.8 or 27.3 for men and women respectively; PHS, 1991) assessed exercise/fitness related behaviors. Finally, a dichotomous indicator of drinking status ("During the past month, have you had at least one drink of any alcoholic beverage?"; yes = 1), the average amount of alcohol consumed per drinking day among drinkers, and a dichotomous indicator of binge drinking (i.e. having consumed five or more drinks on a single occasion during the past month) were used to assess alcohol-related behavior.

Socioeconomic Status Predictors

Educational attainment was categorized into those with less than 12 years of education, those with a high-school degree, those with some college or technical school training beyond high-school, and those with a college degree or more. Employment status was operationalized using a format outlined by Prause and Dooley (1997). Individuals not currently working in paid employment and not currently looking for work were classified as out of the labor force; individuals not currently working but report having looked for a job in the past four weeks were categorized as unemployed. Individuals working in poverty-level wage jobs (i.e. household earnings below the federal poverty standard) or who were involuntarily working part-time due to economic conditions were categorized as inadequately employed, and remaining workers were classified as adequately employed. Both individual and household earnings were measured with single items with common categorization (i.e. < $20,000, $20,000–$40,000, $40,000 to $60,000, $60,000 to $80,000 and > $80,000 per year). Although the association between individual and household earnings was moderately strong ($\tau = 0.63$; $r_s = 0.70$) collinearity diagnostics indicated that inclusion of both measures would not strongly influence parameter estimates or estimated standard errors.

Mediating Mechanisms

The *life strain hypothesis* was assessed using four indicators of environmental or psychological strain. *Low decision latitude* on the job was assessed by averaging responses to the following three items: (1) "My job requires that I learn new things", (2) "On my job, I have very little freedom to decide how I do work", and (3) "My job allows me to make a lot of decisions on my own" ($\alpha = 0.51$). *High psychological demands* on the job was assessed by averaging responses to "I have enough time to get the job done" and "My job requires working very fast without having breaks" ($\alpha = 0.34$). Response categories for both job characteristic constructs were strongly agree = 4, agree somewhat = 3, disagree somewhat = 2, and strongly disagree = 1 with appropriate items recoded. Being employed in an *unsafe neighborhood* was measured by responses to six items (e.g. "How much of a problem is crime in the area" or "How much of a problem is trash and litter"?). Response categories were very serious = 4, somewhat serious = 3, minor problem = 2, and not really a problem = 1 ($\alpha = 0.63$). Each of the employment characteristics was dichotomized based upon a mean split (e.g. below the mean level of decision latitude = 1, above the mean of psychological demand = 1, and individuals responding "very serious" to each neighborhood item were coded 1). Separate indicator variables were constructed for respondents who were legitimately missing employment characteristics measures (i.e. respondents who were not employed or out of the labor force) and employed individuals with missing values to provide more reliable parameter estimates (Orme & Reis, 1990). Finally, a question asking respondents "Including yourself, how many people live in your household?" was used to tap family-related strain.

The *social resources* mediating hypothesis were measured with a series of single item questions. The level and quality of social ties was assessed with five separate items including: (1) "How many close friends do you have?", (2) "How many relatives do you have that you feel close?", (3) "How many of these friends or relatives do you see at least once a month?", (4) "How often do you talk on the telephone with any close relatives or friends – more than once a week, once a week, a few times a month, once a month, or less than once a month?", and (5) "Overall, how satisfied are you with your friendships and relationships with others – very satisfied, somewhat satisfied, somewhat satisfied, or very dissatisfied?"

The *psychological resources* hypothesis was examined by measuring the respondent's level of *depression*, using the Geriatric Depression Scale (Rule, Harvey & Dobbs, 1989; Sheikh & Yesavage, 1986), and feelings of control. Perceived feelings of control were measured by calculating the mean response to four items from Cohen, Kamarck, and Mermelstein (1983) "In the past month, how often have you: (1) felt you were unable to control the important things in your life, (2) felt confident about your ability to handle your personal problems, (3) felt that things were going your way, and (4) felt difficulties piling up so high that you could not overcome them" ? With appropriate items recoded, response categories were almost never = 4, sometimes = 3, often = 2, and very often = 1 ($\alpha = 0.59$).

Finally the medical care hypothesis was assessed with a single item asking the respondent "Do you currently have any kind of health care coverage?". Respondents answering yes were coded 1, otherwise the respondent was coded 0.

Analytic Sequence

Three sets of analyses were undertaken in this study. First, ordinary least squares and logistic regression models were used to describe the association between different socioeconomic indicators and different health behaviors, adjusting only for age and marital status. Next, regression techniques were used to estimate the unique association of multiple measures of SES (typifying the progression of socioeconomic status throughout the life course; Marks & Shinberg, 1997) and health behavior outcomes. That is, each health behavior was regressed on (in the order of the step-wise procedure) race/ethnicity, educational attainment, employment status, individual income, and household income. Finally, following the established procedures for testing mediating hypotheses (Baron & Kenny, 1986), separate models tested the ability of proposed mediators to explain identified associations between socioeconomic status and health behaviors. Given previous empirical evidence indicating differential effects of SES on health behaviors by gender (Marmot, Kogevinas & Elston, 1987), gender separate models were examined. Factors used in the sampling strategy were controlled in most analyses, and the pattern of findings for both weighted and unweighted analyses were similar; consequently, unweighted results are reported in all the regression tables (Winship & Radbill, 1994).

RESULTS

Descriptive Statistics

Respondents in this sample were predominantly middle-aged, non-married, of white non-Hispanic ethnicity with some college or technical training. Approximately 60% of the respondents in this sample were employed either in an adequate or sub-adequate job, and the largest percent of individuals reported individual and household earnings ranging from \$20,000 to \$40,000. In contrast to men, women had lower levels of education ($p \leq 0.05$), lower individual incomes ($p \leq 0.000$) (if they were employed), lower household earnings ($p \leq 0.000$), and were less likely to be adequately employed ($p \leq 0.000$).

According to these data 18.8% of Californians smoke, mirroring prevalence estimates of 19.2% by the Behavioral Risk Factor Surveillance System (BRFSS; Centers for Disease Control [CDC], 1999b). Similar to national estimates the prevalence of smoking in California among men is greater than that of women (21.3% vs. 16.3% respectively; $p \leq 0.001$; CDC, 1999b). Among smokers, men smoke more cigarettes per day in contrast to women both in California (14 vs. 10 cigarettes per day respectively; $p \leq 0.05$) and across the nation (CDC, 1999b).

Estimates from the California Work and Health Survey indicate that 22.5% of Californians did not engage in any physical activity last month (i.e. Sedentary Lifestyle), paralleling the 25.5% estimate for Californians from the Behavioral Risk Factor Surveillance System (CDC, 1999b). Also consistent with BRFSS estimates indicating that women are more likely to be inactive than men, CWHS data indicated that the odds of being sedentary were 56% higher for women than men (CI, 1.25–1.96). These data further revealed however that among active individuals women were nearly 59% more likely to engage in regular aerobic activity (CI, 1.28–1.96) since the primary mode of

Table 1. Descriptive Statistics for all Health Behavior and Socioeconomic Status Variables

	Combined	Women	Men
Health Behaviors			
Smoking Status***	18.8%	16.3%	21.3%
Average Cigarettes/ Day*	12.35	10.24	13.97
	(9.56)	(8.01)	(10.33)
Sedentary Lifestyle***	22.5%	25.8%	19.1%
Aerobic Activity 3X/ week***	26.8%	32.5%	21.1%
Obese	25.9%	23.7%	28.2%
Reports Drinking in past month***	51.8%	44.1%	59.7%
Drinks per Drinking day***	2.44	1.82	2.91
	(2.63)	(1.26)	(3.23)
Binge Drinking***	29.5%	18.1%	38.1%
Sociodemographic Characteristics			
Age	42.07	43.32	40.81
	(17.46)	(17.72)	(17.11)
Marital Status (married = 1)	43.1%	43.8%	42.5%
Non-Hispanic White	65.8%	65.2%	66.3%
Black	8.0%	8.1%	7.9%
Asian	9.5%	9.7%	9.3%
Hispanic	16.7%	16.9%	16.5%
*Education**			
Less than High School	12.5%	12.8%	12.1%
High School	24.7%	25.1%	24.3%
Some College	32.6%	33.8%	31.3%
College Graduate	30.3%	28.2%	32.4%
*Employment Status****			
Adequately Employed	50.7%	42.4%	59.2%
Inadequately Employed	10.8%	10.6%	10.1%
Unemployed	4.2%	4.1%	4.4%
Out of the Labor Force	34.2%	43.0%	25.4%
*Individual Income****			
Missing	48.5%	56.5%	40.3%
Less than $20,000	16.9%	17.1%	16.7%
$20,000 to $40,000	17.0%	15.0%	19.1%
$40,000 to $60,000	9.3%	5.9%	12.7%
$60,000 to $80,000	3.9%	2.3%	5.5%
Greater than $80,000	4.4%	3.1%	5.7%
*Household Earnings****			
Missing	13.9%	15.7%	12.2%
Less than $20,000	20.1%	23.8%	16.4%
$20,000 to $40,000	24.9%	23.6%	26.3%
$40,000 to $60,000	17.4%	15.9%	19.0%
$60,000 to $80,000	9.7%	9.0%	10.5%
Greater than $80,000	13.9%	12.1%	15.8%

Source: California Work and Health Survey, 1998.
Note: Analyses based on weighted data. Chi-square and t-tests used to assess gender differences.
* $p \leq 0.05$ ** $p \leq 0.01$ *** $p \leq 0.001$ (two-tailed).

aerobic activity reported by respondents was walking. The estimated prevalence of obesity among Californians is lower in these data in contrast to BRFSS estimates (26% vs. 31% respectively; CDC, 1999b), and there is no difference in the prevalence of obesity between women and men in California.

In contrast to 1997 Behavioral Risk Factor Surveillance System estimates (CDC, 1999b), the California Work and Health survey sample had more people that abstained from alcohol in the past month (i.e. 61.4% vs. 51.8% respectively reported consuming any alcohol in the past month). Among drinkers, an average of $2\frac{1}{2}$ drinks were consumed per drinking occasion, and over a quarter of the population reported consuming five or more drinks per setting in the past month which is slightly higher than estimates of binge drinking among Californians obtained by the BRFSS (i.e. 24.9%; CDC, 1999b). Consistent with national patterns the odds of abstaining from alcohol in the last month were greater for women than men (0.54: 1.00; CI, 0.45–0.65), while women also drank fewer drinks per occasion (1.8 vs. 2.9; $p \leq 0.000$) and had lower odds of reporting binge drinking (0.35: 1.00; CI, 0.25–0.48).

Socioeconomic Status and Health Behaviors

Smoking-Related Behavior

Education was the only aspect of socioeconomic status in this study to exert a clear linear affect on smoking status, but this pattern was found only in men. In contrast to college educated men, the odds of being a current smoker were higher for men at each of the lower levels of educational attainment. However, the clear monotonically ordered disadvantage of less education did not emerge until other aspects of SES (i.e. employment status, individual income and household income) were adjusted. After adjusting for the effects of education, men whose household earnings were less than $20,000 were also over $2\frac{1}{2}$ times more likely to be a smoker in contrast to men in the highest category of household earnings. Yet in terms of individual income, men in the $60,000 to $80,000 income category were over four times more likely to be a smoker in contrast to those in the highest individual income category. The odds of being a smoker were 50% lower for men who were out of the labor force in contrast to individuals in adequate jobs. Among women, the only indicator of SES found to be associated with smoking status was education: high school educated women or women with a GED were nearly two times more likely to be a smoker in contrast to college educated women.

Although different aspects of socioeconomic status appears to partially condition who becomes a smoker, these analyses did not provide any robust evidence that the number of cigarettes smoked per day differed by any aspect of SES. Although it is not an indicator of SES, ethnicity was found to be associated with the number of cigarettes smoked per day after adjusting for multiple dimensions of SES; people of color were found to smoke fewer

Table 2a. Estimated Associations between Sociodemographic Characteristics and Smoking-Related Behavior.

	Smoking Status[a]			
	Women		Men	
	Model 1[b]	Model 2[b]	Model 1[b]	Model 2[b]
Race/Ethnicity				
White	Reference	Reference	Reference	Reference
Black	0.51*	0.48*	1.14	0.99
Asian	0.41*	0.47+	0.94	1.08
Hispanic	0.59+	0.44*	1.00	0.70
Education				
Less than H.S.	1.20	1.75	2.30**	2.92**
H.S. or G.E.D.	1.92**	2.02**	2.39***	2.79***
Some College	1.40	1.40	1.79**	2.00**
College Graduate	Reference	Reference	Reference	Reference
Employment Status				
Adequately Employed	Reference	Reference	Reference	Reference
Inadequately Employed	1.44	1.53	1.31	1.04
Unemployed	0.39	0.58	1.24	0.57
Out of the Labor Force	1.37	1.64	1.07	0.50*
Individual Income[c]				
Less than $20,000	0.68	0.48	2.35+	0.91
$20,000 to $40,000	0.61	0.54	1.71	1.02
$40,000 to $60,000	0.42	0.41	1.56	1.44
$60,000 to $80,000	0.002	0.003	3.85**	4.31**
Greater than $80,000	Reference	Reference	Reference	Reference
Household Income[c]				
Less than $20,000	1.17	1.24	2.46**	2.65**
$20,000 to $40,000	1.09	1.14	1.54	1.55
$40,000 to $60,000	1.07	1.18	1.12	1.00
$60,000 to $80,000	0.47	0.70	1.44	0.94
Greater than $80,000	Reference	Reference	Reference	Reference
Log Likelihood		706.04		812.93

Source: California Work and Health Survey, 1998.

+ $p \leq 0.10$ * $p \leq 0.05$ ** $p \leq 0.01$ *** $p \leq 0.001$ (two-tailed).

[a] Dichotomous Outcome; Estimated Odds Ratio reported.

[b] Model controls for the effects of age and marital status.

[c] Models estimated using flags for missing values (Orme & Reis, 1990).

Table 2b. Estimated Associations between Sociodemographic Characteristics and Smoking-Related Behavior.

	Cigarettes Smoked per Day[a]			
	Women		Men	
	Model 1[b]	Model 2[b]	Model 1[b]	Model 2[b]
Race/Ethnicity				
White	Reference	Reference	Reference	Reference
Black	0.017	–0.034	–0.151*	–0.175**
Asian	–0.177*	–0.125	–0.195**	–0.167*
Hispanic	–0.294***	–0.360***	–0.156*	–0.173*
Education				
Less than H.S.	–0.019	0.137	0.031	–0.079
H.S. or G.E.D.	–0.022	–0.073	0.014	–0.020
Some College	0.198+	0.209+	0.068	0.103
College Graduate	Reference	Reference	Reference	Reference
Employment Status				
Adequately Employed	Reference	Reference	Reference	Reference
Inadequately Employed	–0.052	–0.011	–0.095	–0.063
Unemployed	–0.047	–0.058	0.059	0.099
Out of the Labor Force	0.035	–0.049	–0.036	–0.044
Individual Income[c]				
Less than $20,000	0.042	0.043	–0.225	–0.277
$20,000 to $40,000	0.177	0.120	–0.205	–0.171
$40,000 to $60,000	0.099	0.037	–0.146	–0.206
$60,000 to $80,000	*e*	*e*	–0.221	–0.208
Greater than $80,000	Reference	Reference	Reference	Reference
Household Income[c]				
Less than $20,000	–0.045	–0.043	0.092	0.214+
$20,000 to $40,000	0.160	0.097	–0.131	–0.132
$40,000 to $60,000	0.027	0.018	0.062	0.084
$60,000 to $80,000	–0.030	0.095	–0.076	–0.052
Greater than $80,000	Reference	Reference	Reference	Reference
Constant		2.48		11.67**

Source: California Work and Health Survey, 1998.
+ $p \leq 0.10$ * $p \leq 0.05$ ** $p \leq 0.01$ *** $p \leq 0.001$ (two-tailed).
[a] Continuous outcome; Standardized Beta reported
[b] Model controls for the effects of age and marital status
[c] Models estimated using flags for missing values (Orme & Reis, 1990)
e: No smokers in this individual income category.

cigarettes per day than whites (Hispanics only differed from whites among women).

Table 3a. Estimated Associations between Sociodemographic Characteristics and Fitness-Related Outcomes.

	Sedentary[a]			
	Women		Men	
	Model 1[b]	Model 2[b]	Model 1[b]	Model 2[b]
Race/Ethnicity				
White	Reference	Reference	Reference	Reference
Black	1.38	1.32	1.17	1.08
Asian	1.16	1.04	0.75	0.78
Hispanic	3.25***	1.81*	4.15***	2.86***
Education				
Less than H.S.	4.64***	2.28**	3.33***	1.47
H.S. or G.E.D.	2.55***	2.03**	2.13**	1.49
Some College	1.25	1.05	0.98	0.75
College Graduate	Reference	Reference	Reference	Reference
Employment Status				
Adequately Employed	Reference	Reference	Reference	Reference
Inadequately Employed	2.21**	1.18	1.82*	0.95
Unemployed	2.63**	3.55*	1.75	2.09
Out of the Labor Force	2.02***	2.35*	1.19	1.24
Individual Income[c]				
Less than $20,000	1.91	0.76	2.71*	1.04
$20,000 to $40,000	0.79	0.63	2.06	1.02
$40,000 to $60,000	0.81	1.00	1.18	1.26
$60,000 to $80,000	0.78	0.90	0.86	1.12
Greater than $80,000	Reference	Reference	Reference	Reference
Household Income[c]				
Less than $20,000	3.96***	1.92+	3.49***	2.59*
$20,000 to $40,000	1.93*	1.28	2.15*	2.00
$40,000 to $60,000	0.94	0.78	1.61	1.43
$60,000 to $80,000	0.90	0.90	0.74	0.65
Greater than $80,000	Reference	Reference	Reference	Reference
Log Likelihood/Constant		885.04		694.26

Source: California Work and Health Survey, 1998.
+ $p \leq 0.10$ * $p \leq 0.05$ ** $p \leq 0.01$ *** $p \leq 0.001$ (two-tailed).
[a] Dichotomous Outcome; Estimated Odds Ratio reported
[b] Model controls for the effects of age and marital status.
[c] Modes estimated using flags for missing values (Orme & Reis, 1990)

Fitness-related Behavior

Education, employment, and earnings disadvantage were all found to be associated with increased odds of being sedentary in the simple models (i.e.

Table 3b. Estimated Associations between Sociodemographic Characteristics and Fitness-Related Outcomes.

	Aerobic Activity ≥ 3 Times Per Week[a]			
	Women		Men	
	Model 1[b]	Model 2[b]	Model 1[b]	Model 2[b]
Race/Ethnicity				
White	Reference	Reference	Reference	Reference
Black	1.08	1.09	0.87	0.86
Asian	0.97	0.95	1.01	0.86
Hispanic	0.610*	0.67	0.51*	0.63
Education				
Less than H.S.	0.63 +	0.62	0.30***	0.30**
H.S. or G.E.D.	0.95	0.83	0.70 +	0.60*
Some College	0.980	0.90	0.62*	0.60*
College Graduate	Reference	Reference	Reference	Reference
Employment Status				
Adequately Employed	Reference	Reference	Reference	Reference
Inadequately Employed	0.67	0.52*	0.86	0.89
Unemployed	0.76	0.94	1.28	1.16
Out of the Labor Force	0.95	1.01	1.06	0.99
Individual Income[c]				
Less than $20,000	2.42 +	2.71 +	0.90	1.40
$20,000 to $40,000	1.52	1.11	0.70	0.84
$40,000 to $60,000	2.83 +	2.06	0.85	0.86
$60,000 to $80,000	1.96	1.20	0.88	0.87
Greater than $80,000	Reference	Reference	Reference	Reference
Household Income[c]				
Less than $20,000	1.06	1.25	0.83	1.14
$20,000 to $40,000	1.16	1.42	0.90	1.14
$40,000 to $60,000	1.30	1.23	0.90	1.09
$60,000 to $80,000	1.63	1.69	0.87	1.00
Greater than $80,000	Reference	Reference	Reference	Reference
Log Likelihood/Constant		1039.34		834.51

Source: California Work and Health Survey, 1998.
+ $p \leq 0.10$ * $p \leq 0.05$ ** $p \leq 0.01$ *** $p \leq 0.001$ (two-tailed).
[a] Dichotomous Outcome; Estimated Odds Ratio reported
[b] Model controls for the effects of age and marital status
[c] Modes estimated using flags for missing values (Orme & Reis, 1990)

Model 1) assessing fitness-related behavior adjusting only for age and marital status. Among women the disadvantage of low education, not being employed and low earnings (trend level) on being sedentary remained robust even after

Table 3c. Estimated Associations between Sociodemographic Characteristics and Fitness-Related Outcomes.

	Obesity[a]			
	Women		Men	
	Model 1[b]	Model 2[b]	Model 1[b]	Model 2[b]
Race/Ethnicity				
White	Reference	Reference	Reference	Reference
Black	1.87**	1.89**	1.19	1.25
Asian	0.36**	0.37**	0.27***	0.31***
Hispanic	2.76***	2.15**	1.18	1.09
Education				
Less than H.S.	2.67***	1.29	1.56+	1.33
H.S. or G.E.D.	1.50+	1.07	1.97***	1.75*
Some College	1.46+	1.14	1.53*	1.45+
College Graduate	Reference	Reference	Reference	Reference
Employment Status				
Adequately Employed	Reference	Reference	Reference	Reference
Inadequately Employed	1.80*	1.38	1.00	0.94
Unemployed	1.50	1.68	0.56	0.93
Out of the Labor Force	1.12	1.00	0.68+	1.03
Individual Income[c]				
Less than $20,000	2.36	0.65	0.91	0.95
$20,000 to $40,000	2.00	0.93	0.80	0.94
$40,000 to $60,000	1.91	1.03	0.78	0.98
$60,000 to $80,000	0.31	0.19	0.79	0.98
Greater than $80,000	Reference	Reference	Reference	Reference
Household Income[c]				
Less than $20,000	3.22***	2.34*	0.88	0.83
$20,000 to $40,000	2.48**	1.97+	0.85	0.71
$40,000 to $60,000	1.77+	1.65	0.80	0.69
$60,000 to $80,000	1.16	1.50	0.78	0.67
Greater than $80,000	Reference	Reference	Reference	Reference
Log Likelihood/Constant		905.86		897.13

Source: California Work and Health Survey, 1998.
+ $p \leq 0.10$ * $p \leq 0.05$ ** $p \leq 0.01$ *** $p \leq 0.001$ (two-tailed).
[a] Dichotomous Outcome; Estimated Odds Ratio reported
[b] Model controls for the effects of age and marital status
[c] Modes estimated using flags for missing values (Orme & Reis, 1990)

adjusting for the effects of multiple indicators of socioeconomic status (i.e. Model 2). The odds of being sedentary were over two-times greater for women with a high school degree or less in contrast to women who graduated from college. Controlling for the effects of education however, women who were out of the labor force or unemployed were 2 to $3\frac{1}{2}$ times more likely to be sedentary in contrast to women who were adequately employed.

Among men the only aspect of SES that was associated with being sedentary was household earnings: the odds of being sedentary were $2\frac{1}{2}$ times greater for men whose household earnings were less than $20,000 in contrast to men in the highest earnings category. Independent of the effects of multiple dimensions of SES, being of Hispanic origin was found to be associated with greater odds of being sedentary among both women and men.

Education was the only aspect of socioeconomic status that was found to be associated with participation in aerobic activity three or more times in the past week, and this relationship was only evident among men. Each level of education was found to be associated with significantly lower odds of participating in regular physical activity. Among women inadequate employment was the only socioeconomic correlate of aerobic activity: the odds of reporting participation in regular aerobic activity last week were 48% lower for those who were underemployed in contrast to adequately employed women. Controlling for education and employment, trend level evidence suggested that women who personally earn less than $20,000 per year were nearly three times more likely to engage in physical activity than women in the highest income category (perhaps because women with low earnings are more likely to walk to work; Ford et al., 1991).

Education, employment and household earnings were all significant correlates of obesity among women in simple models adjusting for only age and marital status. However, only women's household earnings remained independently associated with obesity once each dimension of SES was controlled. These results suggest that low levels of household earnings mediate the association between educational disadvantage and obesity among women. By contrast the odds of being obese remains higher among educationally disadvantaged men in contrast to college graduates in the unique effects model, although the education-obesity association is attenuated somewhat by other aspects of SES.

Alcohol-Related Behavior

Education, employment status, and earnings were all found to be significantly associated with reporting drinking in the past month among women and men after adjusting for age and marital status (Model 1). Among women education and income remained robust correlates of drinking status once the independent effects of all aspects of socioeconomic status were considered simultaneously. Women with less education and lower household earnings were less likely to

Table 4a. Estimated Associations between Sociodemographic Characteristics and Alcohol-Related Behavior.

	Drank last Month[a]			
	Women		Men	
	Model 1[b]	Model 2[b]	Model 1[b]	Model 2[b]
Race/Ethnicity				
White	Reference	Reference	Reference	Reference
Black	0.60*	0.64 +	0.92	1.05
Asian	0.34***	0.33***	0.64*	0.61*
Hispanic	0.30***	0.56*	0.65 +	0.86
Education				
Less than H.S.	0.20***	0.40**	0.55**	0.84
H.S. or G.E.D.	0.45***	0.61*	0.89	1.14
Some College	0.58***	0.71 +	0.87	1.06
College Graduate	Reference	Reference	Reference	Reference
Employment Status				
Adequately Employed	Reference	Reference	Reference	Reference
Inadequately Employed	0.68 +	1.08	0.72	1.49
Unemployed	0.29**	0.50	0.60	0.42*
Out of the Labor Force	0.44***	0.61 +	0.66*	0.42**
Individual Income[c]				
Less than $20,000	0.70	2.40	0.36**	0.55
$20,000 to $40,000	1.32	2.67 +	0.59	0.97
$40,000 to $60,000	1.14	1.72	0.69	1.03
$60,000 to $80,000	1.86	2.34	0.87	0.88
Greater than $80,000	Reference	Reference	Reference	Reference
Household Income[c]				
Less than $20,000	0.19***	0.28***	0.41***	0.52*
$20,000 to $40,000	0.39***	0.44**	0.44***	0.51*
$40,000 to $60,000	0.54*	0.51*	0.60*	0.61 +
$60,000 to $80,000	0.80	0.78	0.92	0.93
Greater than $80,000	Reference	Reference	Reference	Reference
Log Likelihood		1058.71		1031.80

Source: California Work and Health Survey, 1998.
+ $p \leq 0.10$ * $p \leq 0.05$ ** $p \leq 0.01$ *** $p \leq 0.001$ (two-tailed).
[a] Dichotomous Outcome; Estimated Odds Ratio reported
[b] Model controls for the effects of age and marital status
[c] Models estimated using flags for missing values (Orme & Reis, 1990)

report being a drinker in contrast to their counterparts with a college education and those with household earnings greater than $80,000. Interestingly, trend level evidence (i.e. $p \leq 0.10$) suggests that women with individual earnings of

Table 4b. Estimated Associations between Sociodemographic Characteristics and Alcohol-Related Behavior.

	Drinks per Drinking Day[a]			
	Women		Men	
	Model 1[b]	Model 2[b]	Model 1[b]	Model 2[b]
Race/Ethnicity				
White	Reference	Reference	Reference	Reference
Black	0.001	0.000	0.014	0.004
Asian	–0.042	–0.036	–0.069	–0.036
Hispanic	0.026	–0.035	0.041	0.036
Education				
Less than H.S.	0.155**	0.160**	0.068	0.052
H.S. or G.E.D.	0.05	0.057	0.134**	0.120*
Some College	0.160**	0.164**	0.098*	0.064
College Graduate	Reference	Reference	Reference	Reference
Employment Status				
Adequately Employed	Reference	Reference	Reference	Reference
Inadequately Employed	0.045	–0.049	0.024	–0.116*
Unemployed	–0.093 +	–0.087	–0.041	–0.023
Out of the Labor Force	0.084	0.081	0.034	0.064
Individual Income[a]				
Less than $20,000	0.262**	0.259*	0.076	0.171 +
$20,000 to $40,000	0.137	0.172	–0.006	–0.026
$40,000 to $60,000	0.042	0.046	–0.041	–0.055
$60,000 to $80,000	0.072	0.111	–0.034	–0.058
Greater than $80,000	Reference	Reference	Reference	Reference
Household Income[a]				
Less than $20,000	0.07	–0.072	0.017	–0.063
$20,000 to $40,000	–0.008	–0.123 +	0.027	–0.044
$40,000 to $60,000	0.008	–0.031	0.008	0.007
$60,000 to $80,000	–0.057	–0.101 +	0.082	0.101 +
Greater than $80,000	Reference	Reference	Reference	Reference
Constant		2.25***	4.40***	3.85***

Source: California Work and Health Survey, 1998.
+ $p \leq 0.10$ * $p \leq 0.05$ ** $p \leq 0.01$ *** $p \leq 0.001$ (two-tailed).
[a] Continuous Outcome; Standardized Beta reported
[b] Model controls for the effects of age and marital status
[c] Models estimated using flags for missing values (Orme & Reis, 1990)

$20,000 to $40,000 are more likely to report drinking in contrast to the highest paid women. Among men, education was not associated with being a drinker, but those who were not currently working and those with lower levels of

Table 4c. Estimated Associations between Sociodemographic Characteristics and Alcohol-Related Behavior.

	Binge Drinking[a]			
	Women		Men	
	Model 1[b]	Model 2[b]	Model 1[b]	Model 2[b]
Race/Ethnicity				
White	Reference	Reference	Reference	Reference
Black	1.25	1.02	0.86	0.75
Asian	0.41	0.41	0.57+	0.70
Hispanic	2.13+	1.37	1.35	1.20
Education				
Less than H.S.	4.79**	6.69**	2.53**	2.47*
H.S. or G.E.D.	1.00	1.29	2.01**	2.19**
Some College	2.04*	2.52*	1.79*	1.93**
College Graduate	Reference	Reference	Reference	Reference
Employment Status				
Adequately Employed	Reference	Reference	Reference	Reference
Inadequately Employed	1.72	1.18	1.14	0.96
Unemployed	0.69	1.30	1.60	1.85
Out of the Labor Force	1.12	1.51	0.83	1.17
Individual Income[c]				
Less than $20,000	1.47	1.44	1.08	0.40
$20,000 to $40,000	1.08	1.51	1.22	0.51
$40,000 to $60,000	0.28	0.39	0.98	0.47
$60,000 to $80,000	2.21	5.33	1.54	1.03
Greater than $80,000	Reference	Reference	Reference	Reference
Household Income[c]				
Less than $20,000	0.92	0.42	1.51	1.68
$20,000 to $40,000	0.79	0.39+	1.28	1.25
$40,000 to $60,000	0.36+	0.37+	1.91*	2.31*
$60,000 to $80,000	0.63	0.32	1.61	1.75
Greater than $80,000	Reference	Reference	Reference	Reference
Log Likelihood		287.31		565.76

Source: California Work and Health Survey, 1998.
+ $p \leq 0.10$ * $p \leq 0.05$ ** $p \leq 0.01$ *** $p \leq 0.001$ (two-tailed).
[a] Dichotomous Outcome; Estimated Odds Ratio reported
[b] Model controls for the effects of age and marital status
[c] Models estimated using flags for missing values (Orme & Reis, 1990)

household earnings were less likely to report having consumed alcohol in the month prior to the interview.

Education, individual income, and household earnings (trend level) were all found to be independently and uniquely associated with the number of drinks consumed per drinking day among women. Women with less than a high school education and women with some college consumed more alcohol on the days they drank in contrast to women with a college education. Controlling for the effects of education, women in the lowest income category were also found to consume more drinks on drinking days than women in the highest income category. Further, and in contrast to a pattern indicating that disadvantaged women drink more, trend level effects suggest that women in the $20,000 to $40,000 category as well as those in the $60,000 to $80,000 category for household earnings drink less than women in the highest household earnings category.

Among men who drink, results indicated that high school graduates consumed a larger number of alcoholic beverages on a typical drinking occasion than college educated men. Men who worked in inadequate jobs however consumed fewer drinks than adequately employed men, controlling for education and income. A trend level finding suggests that men in the lowest individual income category, controlling for the economic adequacy of the job and education, consumed more drinks per drinking day than men in the highest income category. Another trend level finding however suggested that alcohol consumption patterns did not differ between the most and least affluent in terms of household earnings, but rather differences may exist between the top two earnings categories.

Educational differences in binge drinking are clearly evident among women and men both in the simple models adjusting for age and marital status and in the unique effects models. The odds of reporting an episode of binge drinking in the month prior to the interview was significantly greater for individuals with less than a college degree, although no differences in binge drinking were found between women who were high school graduates and those who were college graduates. Independent of the effects for education, the odds of binge drinking were lower (trend level) for women whose household earnings fell in the $20,000 to $40,000 or the $40,000 to $60,000 category in contrast to women with the highest household earnings. These trend level findings are provocative since the odds of consuming five or more drinks in a single setting in the month prior to the interview was over two times greater for men in the $40,000 to $60,000 category in contrast to men in the highest household earnings category.

Explanations of SES and Health Behavior Associations

The "psychological resources" explanation for the socioeconomic status-health behavior relationship received the most support in these analyses. Differences in sedentary lifestyle by income among men, and differences by employment status and education among women were completely explained by higher levels of depression (analyses not shown). A lower level of perceived control was also found to explain the deleterious associations between low education and sedentary lifestyle among women, as well as the association between not working and sedentary lifestyle. Lower feelings of control slightly attenuated the association between lower levels of education and smoking status among women. Among men, a higher level of depression (and the lack of insurance, at a trend level) was found to explain a small amount of the association between education and smoking status. Finally, some evidence indicated that higher levels depression (trend level), as well as more frequent contact with friends (trend level), accounted for a small amount of the association between education and the number of drinks consumed per drinking day among men.

Among men only, employment stresses were found to partially mediate the associations between education and obesity, as well as part of the association between education and participation in regular aerobic activity the week prior to the interview. That is, working in unsafe working conditions was found to mediate a small amount of the association between education and obesity, while low levels of decision latitude explained part of the association between low education and less aerobic activity. It is important to note that with the exception of SES differences in sedentary lifestyle, only a very small amount of the various associations between different indicators of SES and health behaviors were explained by the proposed mediators in this study.

SUMMARY AND DISCUSSION

In this study the association between SES and different health behaviors was disaggregated by examining the unique and independent effects of separate aspects of SES on a variety of important health behaviors. A clear gradient, or gradient like association was found between education and six of the eight health behaviors examined in this study suggesting that education is indeed a powerful indicator of socioeconomic status. Interestingly however, four of the six education-health behavior gradients were evidenced among men, while the pattern of findings for education across health behaviors was less consistent for women. Indeed only two educational gradients were found among women, one indicating a health behavior disadvantage for women with lower levels of education (i.e. sedentary lifestyle), and one indicating that women with less education were less likely to report having consumed alcohol in the month prior to the interview. These results support previous arguments suggesting that traditional conceptualizations of SES are gendered (McDonough, Williams,

House & Duncan, 1999) since educational disadvantage was associated with poor health behaviors more frequently in men than women.

Although education was the most consistent socioeconomic correlate of health behaviors in this study it is important to note that other dimensions of socioeconomic status frequently exhibited robust independent effects above and beyond the effects of education. Employment status and/or income were found to influence five health behaviors, while controlling for the effects of education supporting the contention that different aspects of social class may influence an individual's health practices in parallel yet distinct ways. For example, in addition to the greater odds of being sedentary among educationally disadvantaged women the odds of reporting being sedentary were over two-times greater for women who were not working in contrast to those working in adequate jobs. Similarly a low level of household earnings increased the estimated odds of being smoker among men, even after adjusting for education. Noteworthy about both of these examples is the magnitude of these net effects; that is, the estimated odds of not working for pay and having low household earnings is comparable to or greater than the apparent deleterious effects of educational disadvantage.

An interesting pattern emerged among two health-risk behaviors once all aspects of socioeconomic status were entered into the model. The odds of being a smoker was over four-times greater for men with personal incomes in the $60,000 to $80,000 category in contrast to men in the highest income category, while men in the lowest personal income categories did not differ from the highest paid men. Binge drinking was over two-times more likely for men with household earnings of $40,000 to $60,000 in contrast to men with the highest household earnings (i.e. greater than $80,000), while the prevalence of binge drinking did not significantly differ between men in the poorest and richest households. Paralleling previous reports, these results suggest that relative economic advantage rather than absolute disadvantage may promote more risk-taking behavior after adjusting for education, particularly among men.

Perplexing reports such as these suggesting that individuals with a lower education, and a higher income are most "at risk" for certain negative health behaviors best justifies the importance of considering multiple aspects of socioeconomic status in health behavior research. It is possible, for example, that men with a low education and a high income may have the financial wherewithal to practice risky behaviors yet, despite their high income, they are not held as accountable for their behavior as their highly educated, well paid counterparts (Makela, 1978). Findings and hypotheses such as these, based upon a multidimensional conceptualization of SES, provide a more targeted and fine-grained guide for future research attempting to identify the causal linkages between an individual's location in the opportunity structure and her/his health-related behaviors.

This project also attempted to explore different linkages that may explain the association between different aspects of socioeconomic status and health behaviors. The psychological resources hypothesis received the most support; the association between low income and being sedentary among men, and the associations between not working, lower levels of education and be sedentary among women were completely explained by higher levels of depression. Similarly, a higher level of depression was found to explain a small amount of the association between education and smoking status and the number of drinks consumed per drinking day among men. Strains at work (e.g. job located in an unsafe neighborhood, low levels of decision latitude) were found to explain *small* amounts of the association between education and obesity and aerobic activity among men. With the notable exception of SES-sedentary lifestyle association however, the pattern of results across the health behaviors considered in this study provided few robust explanations for SES-health behavior linkages.

In general, the findings from this study are consistent with the general model anticipating that different aspects of socioeconomic status can each influence health behaviors in similar and different ways, and that the effects of socioeconomic status on health behavior differ for women and men. Despite this empirical support, it is necessary to note the limitations of this study. First, these analyses were based upon cross-sectional data making causal inferences tenuous. Although it is a legitimate limitation, the similarities in the pattern of results in this study and others using longitudinal designs (e.g. Marks, 1996) suggests that socioeconomic status does in fact influence an individual's participation in different health behaviors. Next employment status was used as a proxy for occupational status in this study, but it is usually not considered an indicator of socioeconomic status. Additional research is necessary examining the influence of occupational status on health behaviors. Occupational status has been argued to be a more attractive and valid indicator of overall SES since survey respondents are frequently reluctant to divulge earnings information, but are very willing to share employment related information (Hauser & Warren, 1996). Finally, the psychometric properties of many of the proposed mediators were less than optimal. Although little evidence of mediation was found for many of the proposed mediators, additional research using better operational indicators of life strain, social support, access to care, and psychological resources is necessary before completely ruling out these explanations.

Notwithstanding these limitations, this study furthers our understanding of the association between socioeconomic status and health behaviors. Most specifically the results from this study highlight the relative importance of education as a single indicator of SES, but the results also demonstrate the important and sometimes perplexing influence of different aspects of social class. As the nation continues its quest to promote the health and wellbeing of the entire population through health promotion and disease prevention objectives, it is very much in the public interest to better our understanding of the complex linkages between SES and health-related behavior. Understanding SES, rather than controlling for it, is the first necessary step in developing health interventions that will benefit all.

ACKNOWLEDGMENTS

The author gratefully acknowledges financial support for this research from the Health Policy Institute of the University of California, San Francisco and a National Institute on Mental Health Postdoctoral Traineeship (MH19958). The author also appreciates the helpful guidance and suggestions from Nadine F. Marks of the University of Wisconsin-Madison and Dave Dooley of the University of California, Irvine.

REFERENCES

Adler, N. E., Boyce, T., Chesney, M. A., Cohen, S., Folkman, S., Kahn, R. L., & Syme, S. L. (1994). Socioeconomic Status and Health: The Challenge of the Gradient. *American Psychologist, 49*, 15–24.

Andersen, R. E., Wadden, T. A., Bartlett, S. J., Zemel, B., Verde, T. J., & Franckowiak, S. C. (1999). Effects of Lifestyle Activity Vs Structured Aerobic Exercise in Obese Women: A Randomized Trial. *Journal of the American Medical Association, 281*, 335–340.

Anderson, N. B., & Armstead, C. A. (1995). Toward Understanding the Association of Socioeconomic Status and Health: A New Challenge for the Biopsychosocial Approach. *Psychosomatic Medicine, 57*, 213–225.

Baron, R. M., & Kenny, D. A. (1986). The Moderator-Mediator Variable Distinction in Social Psychological Research: Conceptual, Strategic, and Statistical Considerations. *Journal of Personality and Social Psychology, 51*, 1173–1182.

Berkman, L. F. (1995). The Role of Social Relations in Health Promotion. *Psychosomatic Medicine, 57*, 245–254.

Blaxter, M. (1990). *Health and Lifestyles*. New York: Routledge.

Bollen, K. A. (1984). Multiple Indicators: Internal Consistency or No Necessary Relationship. *Quality and Quantity, 18*, 377–385.

Bollen, K., & Lennox, R. (1991). Conventional Wisdom on Measurement: A Structural Equation Perspective. *Psychological Bulletin, 110*, 305–314.

Centers for Disease Control (1999a). Neighborhood Safety and the Prevalence of Physical Inactivity: Selected States (1996). *Morbidity and Mortality Weekly Report, 48*, 143–46.

Centers for Disease Control (1999b). Online Prevalence Data Available: http://www2.cdc.gov/nccdphp/brfss/index.asp

Cohen, S., Kamarck, T., & Mermelstein, R. (1983). A global measure of perceived stress. *Journal of Health and Social Behavior, 24*, 385–396.

Conway, T. L., Vickers, R. R., Ward, H. W., & Rahe, R. H. (1981). Occupational Stress and Variation in Cigarette, Coffee, and Alcohol Consumption. *Journal of Health and Social Behavior, 22*, 155–165.

Dannenberg, A. L., Keller, J. B., Wilson, P. W. F., & Castelli, W. P. (1989). Leisure Time Physical Activity in the Framingham Offspring Study: Description, Seasonal Variation, and Risk Factor Correlates. *American Journal of Epidemiology, 129*, 76–88.

Dishman, R. K. (1994). The Measurement Conundrum in Exercise. *Medicine and Science in Sports and Exercise, 26*, 1382–1390.

Doll, R., & Peto, R. (1976). Mortality in relation to smoking: Twenty years' observations on male British Doctors. *British Medical Journal, 2*, 1525–1536.

Dunn, A. L., Marcus, B. H., Kampert, J. B., Garcia, M. E., Kohl, H. W., & Blair, S. N. (1999). Comparison of Lifestyle and Structured Interventions to Increase Physical Activity and Cardiorespiratory Fitness: A Randomized Trial. *Journal of the American Medical Association, 281*, 327–334.

Evans, R. G., Barer, M. L., & Marmor, R. R. (Eds) (1994). *Why Are Some People Healthy and Others Not?: The Determinants of Health of Populations*. New York: Walter de Gruyter.

Feinstein, J. S. (1993). The Relationship Between Socioeconomic Status and Health: A Review of the Literature. *The Milbank Quarterly, 71*, 279–322.

Folsom, A. R., Caspersen, C. J., Taylor, H. L., Jacobs, D. R. J., Luepker, R. V., Gomez-Marin, O., Gillum, R. F., & Blackburn, H. (1985). Leisure Time Physical Activity and Its Relationship to Coronary Risk Factors in the a Population Based Sample: The Minnesota Heart Survey. *American Journal of Epidemiology, 121*, 570–579.

Ford, E. S., Merritt, R. K., Health, G. W., Powell, K. E., Washburn, R. A., Kriska, A., & Haile, G. (1991). Physical Activity Behaviors in Lower and Higher Socioeconomic Status Populations. *American Journal of Epidemiology, 133*, 1246–1256.

Gochman, D. S. (Ed.) (1997). *Handbook of Health Behavior Research: Demography, Development and Diversity*. New York: Plenum Press.

Grant, B. F., Harford, T. C., Dawson, D. A., Chou, P., Dufour, M., & Pickering, R. (1994). Prevalence of DSM-IV Alcohol Abuse and Dependence. *Alcohol Health and Research World, 18:*, 243–248.

Grzywacz, J. G., & Marks, N. F. (2000). Family, Work, Work-Family Spillover and Problem Drinking During Midlife. *Journal of Marriage and the Family, 62*, 336–348

Haan, M. N., Kaplan, G. A., & Camacho, T. (1989). Socioeconomic Status and Health: Old Observations and New Thoughts. In: J. P. Bunker, D. S. Gomby & B. H. Kehrer (Eds), *Pathways to Health: The Role of Social Factors* (pp. 76–117). Menlo Park, CA: Henry Kaiser Family Foundation.

Hauser, R. M., & Warren, J. R. (1996). *Socioeconomic Indexes for Occupations: A Review, Update, and Critique*. CDE Working Paper No. 96–01. University of Wisconsin-Madison: Center for Demography and Ecology.

Hilton, M. E. (1990a). Demographic characteristics and the frequency of heavy drinking as predictors of self-reported drinking problems. In: W. B. Clark & M. E. Hilton (Eds), *Alcohol in America: Drinking Practices and Problems* (pp. 194–212). Albany, NY: State University of New York Press.

Hilton, M. E. (1990b). The Demographic Distribution of Drinking pattersn in 1984. In: W. B. Clark & M.l E. Hilton (Eds), *Alcohol in America: Drinking Practices and Problems* (pp. 73–86). Albany, NY: State University of New York Press.

Hilton, M. E. (1990c). The Demographic Distribution of Drinking pattersn in 1984. In: W. B. Clark & M. E. Hilton (Eds), *Alcohol in America: Drinking Practices and Problems* (pp. 87–104). Albany, NY: State University of New York Press.

Kahn, H. S.,Williamson, D. F., & Stevens, J. A. (1991). Race and Weight Change in U.S. Women: The Roles of Socioeconomic and Marital Status. *American Journal of Public Health, 81*, 319–323.

Kessler, R. C. (1982). Disaggregation of the Relationship Between Socioeconomic Status and Psychological Distress. *American Sociological Review, 47*, 752–764.

Kohn, R., Dohrenwend, B. P., & Mirotznik, J. (1998). Epidemiological Findings on Selected Psychiatric Disorders in the General Population. In: B. P. Dohrenwend (Ed.), *Adversity, Stress, and Psychopathology* (pp. 235–284). New York: Oxford University Press.

Krick, J. P., & Sobal, J. (1990). Relationships Between Health Protective Behaviors. *Journal of Community Health, 15*, 19–34.

Krieger, N., Williams, D. R., & Moss, N. E. (1997). Measuring Social Class in U.S. Public Health Research: Concepts, Methodologies, and Guidelines. *Annual Review of Public Health, 18*, 341–378.

Lantz, P. M., House, J. S., Lepkowski, J. M., Williams, D. R., Mero, R. P., & Chen, J. (1998). Socioeconomic Factors, Health Behaviors, and Mortality: Results From a Nationally Representative Prospective Study of U.S. Adults. *Journal of the American Medical Association, 279*, 1703–1708.

Makela, K. (1978). Level of Consumption and Social Consequences of Drinking. In: Y. Israel, F. Glaser, R. Popham, W. Schmidt, H. Kalant & R. Smart (Eds), *Research Advances in Alcohol and Drug Problems* (Vol. 4, pp. 303–348). New York: Plenum Press.

Marks, N. F. (1996). Socioeconomic Status, Gender, and Health at Midlife: Evidence From the Wisconsin Longitudinal Study. In: J. J. Kronenfeld (Ed.), *Research in the Sociology of Health Care: Health Care Delivery System Changes: New Roles for Providers, Insurers, and Patients* (vol. 13, pp. 135–52). Greenwich, CT: JAI Press.

Marks, N. F., & Shinberg, D. S. (1997). Socioeconomic Differences in Hysterectomy: The Wisconsin Longitudinal Study. *American Journal of Public Health, 87*(9), 1507–1514.

Marks, N. F., & Shinberg, D. S. (1998). Socioeconomic Status Differences in Hormone Therapy. *American Journal of Epidemiology, 148*, 581–593.

Marmot, M. G., Kogevinas, M., & Elston, M. A. (1987). Social/Economic Status and Disease. *Annual Review of Public Health, 8*, 111–135.

Marmot, M. G., Smith, G. D., Stansfeld, S., Patel, C., North, F., Head, J., White, I., Brunner, E., & Feeney, A. (1991). Health Inequalities Among British Civil Servants: The Whitehall II Study. *Lancet, 337*, 1387–1393.

Marmot, M., Ryff, C. D., Bumpass, L. L., Shipley, M., & Marks, N. F. (1997). Social Inequalities in Health: Next Questions and Converging Evidence. *Social Science and Medicine, 44*, 901–910.

Matthews, K. A., Kelsey, S. F., Meilahn, E. N., Kuller, L. H., & Wing, R. R. (1989). Educational Attainment and Behavioral and Biologic Risk Factors for Coronary Heart Disease in Middle-Aged Women. *American Journal of Epidemiology, 129*, 1132–1144.

McDonough, P., Williams, D. R., House, J. S., & Duncan, G. J. (1999). Gender and the Socioeconomic Gradient in Mortality. *Journal of Health and Social Behavior, 40*, 17–31.

McLeod, J. D., & Kessler, R. C. (1990). Socioeconomic Status Differences in Vulnerability to Undesirable Life Events. *Journal of Health and Social Behavior, 31*, 162–172.

Mechanic, D. (1989). Socioeconomic Status and Health: An Examination of Underlying Processes. In: J. P. Bunker, D. S. Gomby & B. H. Kehrer (Eds), *Pathways to Health: The Role of Social Factors* (pp. 9–26). Menlo Park, CA: Henry Kaiser Family Foundation.

Millar, W. J., & Wigle, D. T. (1986). Socioeconomic Disparities in Risk Factors for Cardiovascular Disease. *Canadian Medical Association Journal, 134*, 127–132.

National Center for Health Statistics (1993). *Health promotion and disease prevention: United States, 1990*. Hyattsville, MD.

National Center for Health Statistics (1998). *Health, United States, 1998 With Socioeconomic Status and Health Chartbook*. Hyattsville, MD.

Orme, J. G., & Reis, J. (1991). Multiple Regression With Missing Data. *Journal of Social Service Research, 15*, 61–91.

Pocock, S. J., Shaper, A. G., Cook, D. G., Phillips, A. N., & Walker, M. (1987). Social Class Differences in Ischaemic Heart Disease in British Men. *Lancet, 2*, 197–201.

Pol, L. G., & Thomas, R. K. (1997). Demographic Change and Health Behavior. In: D. S. Gochman (Ed.), *Handbook of Health Behavior Research: Demography, Development, and Diversity* (Vol. 3, pp. 3–21). New York: Plenum Press.

Prause, JoA., & Dooley, D. (1997). Effect of Underemployment on School-Leavers' Self-Esteem. *Journal of Adolescence, 20*, 243–260.

Public Health Service (1991). *Healthy People 2000: National Health Promotion and Disease Prevention Objectives: Summary*, Washington D.C.

Rule, B. G., Harvey, H. Z., & Dobbs, A. R. (1989). Reliability of the Geriatric Depression Scale for Younger Adults. *Clinical Gerontology, 9*, 37–43.

Sheikh, J. I., & Yesavage, J. A. (1986). Geriatric Depression Scale (GDS): Recent Evidence and Development of a Shorter Version. *Clinical Gerontology, 5*, 165–173.

Slater, C. H., Lorimor, R. J., & Lairson, D. R. (1985). The Independent Contributions of Socioeconomic Status and Health Practices to Health Status. *Preventive Medicine, 14*, 372–378.

Sobal, J., & Stunkard, A. J. (1989). Socioeconomic Status and Obesity: A Review of the Literature. *Psychological Bulletin, 105*, 260–275.

Strickland, B. R. (1978). Internal-External Expectancies and Health Related Behaviors. *Journal of Consulting and Clinical Psychology, 46*, 1192–1211.

U.S. Department of Health and Human Services (1997). *Ninth Special Report to the U.S. Congress on Alcohol and Health*. Rochville, MD: author.

Williams, D. R. (1990). Socioeconomic Differentials in Health: A Review and Redirection. *Social Psychology Quarterly, 53*, 81–99.

Winship, C, & Radbill, L. (1994). Sampling Weights and Regression Analysis. *Sociological Methods and Research, 23*, 230–263.

Wray, L. A., Herzog, A. R., Willis, R. J., & Wallace, R. B. (1998). The Impact of Education and Heart Attack on Smoking Cessation Among Middle-Aged Adults. *Journal of Health and Social Behavior, 39*, 271–294.

RACE, SOCIOECONOMIC STATUS, AND RISK OF A CATASTROPHIC HEALTH CONDITION IN LATER LIFE

Nancy G. Kutner and Donna Brogan

ABSTRACT

Socioeconomic status (SES) is a fundamental social cause of disease. The association of SES with chronic renal failure has received limited study. Older persons and blacks are at significantly increased risk for this health condition, which is also known as end-stage renal disease (ESRD). Among ESRD cases and controls aged 56–88, we investigated whether SES was predictive of ESRD, independent of subjects' race, sex, age, and history of diabetes. Our SES measure combined information about individuals' education and occupation. In race-specific logistic regression models, there was a trend for higher SES to be protective against ESRD. In a combined model, controlling for the effects of race, sex, age, and diabetes, the SES effect was statistically significant. Diabetes and hypertension, important ESRD risk factors, are also related to SES. The adverse "effect" of black race on a health outcome like ESRD may be expressed preferentially among economically disadvantaged persons. Improved understanding of race, SES, and age vis-à-vis a chronic condition such as ESRD requires studies in which cohort changes in health, resources, access to care, and service utilization are monitored longitudinally and analyzed using a life course perspective.

Health, Illness, and Use of Care: The Impact of Social Factors, Volume 18, pages 151–165.

ISBN: 0-7623-0740-4

INTRODUCTION

Lower socioeconomic status (SES) places individuals at increased risk for developing chronic as well as infectious diseases (Link & Phelan, 1995). Chronic diseases have largely replaced infectious diseases as the major causes of morbidity, disability and mortality in the United States (U.S.), but the same socioeconomic gradient seems to exist for both types of disease. In this paper we examine the association of SES with a chronic health condition, chronic renal failure, for which black persons are at significantly greater risk than are white persons.

Chronic renal failure, or end-stage renal disease (ESRD), exists when normal kidney function has deteriorated to a point at which the kidneys are no longer capable of maintaining the volume, composition, and distribution of body fluids essential for well-being. ESRD patients' survival depends on receiving renal replacement therapy, either maintenance dialysis treatment or a kidney transplant. Black persons are approximately 12% of the U.S. population but are almost one-third of patients receiving treatment for chronic renal failure. The age-sex adjusted incidence of ESRD among blacks (873 per million) is much higher than among whites (218 per million) in the U.S. (USRDS, 1999). Because of black/white differences in average SES in the U.S., the contributions of SES and race to the incidence of a condition such as ESRD are not easily differentiated.

In addition to higher incidence among blacks than among whites, the rate of treated ESRD is also dramatically higher among older persons. The highest ESRD incidence occurs among persons aged 70–74 years, and the highest prevalence is among persons aged 65–69 years. The mean age of patients newly starting therapy for ESRD is 61.1 years (USRDS, 1999). Development of ESRD in older persons may illustrate the phenomenon that Mutchler and Burr (1991) have described as the convergence in later life of the trajectories of both health and socioeconomic "accumulation" of risks. No previous studies that have examined SES as a risk factor for development of ESRD have focused specifically on older persons.

SES, Race and Health

The ubiquitous and often strong association between SES and health is an especially prominent example of the social patterning of disease. Lower SES is acknowledged by the medical community as probably the most powerful single contributor to premature morbidity and mortality (Williams, 1998). An individual's SES encompasses a constellation of life-chances including education, economic success, employment experience, and access to societal goods and resources. These life-chances are critical in avoiding the risks for disease and minimizing the consequences of disease once it occurs.

Goods and resources with particular relevance for health outcomes include knowledge about factors that affect health, e.g. nutrition, and ability to access medical care. Individuals with higher education and income levels are likely to have better health (e.g. Lantz et al., 1998). The relationship between SES and health does not simply reflect individual behavioral choices, however. There is growing consensus that SES differences reflect the cumulative influence of multiple social environments, including family, school, neighborhood, and cultural context (Blane, 1995; Lantz et al., 1998; Williams, 1998).

A number of morbidity and mortality patterns differ sharply for blacks and whites. In addition to their increased risk of ESRD, black Americans have a higher overall incidence of cancer, diabetes, and asthma. Treatment and survival patterns also differ. Blacks are less likely than whites to receive curative surgery for early-stage lung, colon, or breast cancer and are less likely to undergo diagnostic evaluation for symptoms suggestive of coronary artery disease (Freeman & Payne, 2000). Black Americans have lower life expectancy than do whites, with a differential in heart disease being the major contributor (Pappas, 1994). African-Americans are twice as likely as whites to die of cardiovascular disease.

The lower average SES of black persons compared to white persons in the U.S. has been considered a major reason for excess morbidity and mortality among blacks in this country (e.g. Navarro, 1990). Controlling for differences in the SES of blacks and whites does not eliminate race disparities in health, however. Blacks and whites with similar economic and health insurance status often seem to receive different treatments and experience different health outcomes. Adjusting for socioeconomic factors substantially reduces black/white differences in some, but not in all, health outcomes (Krieger et al., 1993). For example, kidney transplantation, compared to maintenance dialysis, is associated with better patient survival and quality of life (Wolfe et al., 1999; Evans et al., 1985), but black ESRD patients are less likely than white patients to be referred for, or to undergo, transplantation, even after adjustment for patients' gender, primary diagnosis, coexisting disease status, sociodemographic characteristics, and treatment preferences (Ayanian et al., 1999; Kutner, 1994). Freeman and Payne (2000: 1046), without specifically alleging that racism exists, suggest that there is often a subtle form of racial bias on the part of medical care providers which contributes to phenomena such as black persons' lower rate of referral for preferred treatment options such as kidney transplantation. It is likely that patient experiences associated with both race/ethnicity and socioeconomic position contribute to racial/ethnic and socioeconomic disparities in health outcomes (Krieger et al., 1997).

Research on the interrelation of race, SES and health is complicated by the fact that imprecise measurement of both race and SES may result in misclassifying individuals in studies that attempt to delineate health outcomes related to these two categories. Misclassification of race is likely to be small in studies in which subjects self-report. Measurement of SES presents more challenges. Level of formal educational achievement and prestige associated with particular occupations may not have the same meaning across racial groups. Income provides a more standard SES measure for comparisons across groups. When subjects are older, however, current income may be limited to retirement and/or disability sources, providing limited differentiation among individuals. There is also the possibility that individuals who have begun a downward health trajectory before a health condition is diagnosed have already experienced income decline, i.e. poor health may be responsible for reduced income rather than vice versa (e.g. Liao et al., 1999).

In addition to the issue of identifying an appropriate individual-level measure of SES, the researcher may need to consider whether it is appropriate to classify subjects by a census block-group measure of SES. Measures that are based on subjects' area of residence may dilute estimates of the effects of class position compared to measures based on individual-level social class data. Residential measures may be the only available ones, however, and these measures may provide an SES classification that yields results comparable to use of an individual or household measure (Krieger, 1991, 1992).

Previous research is limited, but existing studies consistently point to low SES as a significant risk factor for developing chronic renal failure. Investigators have for the most part relied on estimates of subjects' SES derived from the median income of the census tract in which they live and have focused on one gender and/or on younger age samples. Only one study has investigated the contribution of SES to the risk of developing ESRD in both women and men and has measured SES at the individual level; that study did not include persons aged 65+, however, who are the fastest growing component of the ESRD population. ESRD is both catastrophic for individuals and very expensive for society, with the annual costs of treatment rising steadily with patients' age. The interrelation of race, SES, and aging with this disease outcome is therefore an important area of study.

DATA AND METHODS

Samples

The sampling frame for cases was obtained from the local ESRD Network census of patients 60 years or older receiving chronic renal dialysis treatment in Georgia. (ESRD Networks are administrative units with oversight for the ESRD Program of Medicare.) This sampling frame was stratified by race/sex (black males, black females, white males, white females). Each stratum was then sorted by dialysis facility and by patient age within facility, and a systematic random sample was selected. A total of 326 patients were interviewed who had complete education/occupation information. The age range of cases was 60–87.

Interviewed patients were compared to the total ESRD Network census of patients aged 60+ on age, dialysis modality, primary ESRD diagnosis, and length of time on dialysis. In all race/sex strata, the interviewed sample of patients aged 60+ was comparable to the population of patients aged 60+ on these variables, with *p*-values for the chi-square tests and z-tests ranging from 0.24 to 0.99 and most *p*-values being above 0.70.

A sampling frame for controls was obtained from older Georgia residents who had participated in a prior household health survey in the state (Shulman, Martinez, Brogan, Carr & Miles, 1986). This sampling frame was also stratified by race/sex, and each stratum was sorted by geographic area of residence and by age within geographic area. It was necessary to include a small number of controls aged 56–59 in order to match controls with cases on all the desired criteria. In each area/age category, within stratum sampling was conducted so that each race/sex control sample had the same distribution on geographic area and age as did the dialysis population of that race/sex. A total of 334 controls were interviewed who had complete education/occupation information. The age range of controls was 56–88.

Table 1 provides an overview of our study design and of designs used in the previous studies that have investigated race, SES, and risk of ESRD.

Measures

Sociodemographic and health status data were obtained during personal interviews with cases and controls, conducted by trained professional interviewers. Information from the local ESRD Network census was used to verify primary cause of kidney failure among interviewed cases. All subjects

Table 1. Overview of Research Investigating Race, SES, and ESRD Risk

Data Collection Period & Location	Sample Characteristics	No. of Cases	No. of Controls	SES Measure(s)
1980–85: MD, excluding 5 metro DC counties[a]	Males and Females, aged 18+	534	4,427	% completed high school & % with household income ≥ $10,000 in zip code area of residence
1983–88: All U.S. counties[b]	Males and Females, aged < 60	80,172	—	Race-specific average per capita income of patient's county
1982–88: NY State[c]	Males and Females, aged 24+	9,390	—	Race-specific median family income of patient's zip code
1991: MD, VA, West VA, DC[d]	Males and Females, aged 20–64	716	361	Individual's years of education & household annual income
1973–90: 18 U.S. cities[e]	Males only, aged 35–73	814	331,730	Family income
1988: GA[f]	Males and Females, aged 56+	326	334	Individual's education & occupation

[a] Whittle et al. (1991); Brancati et al. (1992).
[b] Young et al. (1994).
[c] Byrne, Nedelman and Luke (1994).
[d] Perneger, Whelton and Klag (1995).
[e] Klag et al. (1996, 1997).
[f] Data reported in this paper.

were asked if they had diabetes or high blood sugar or had taken medication for diabetes during the past six months.

Subjects reported the number of years of school they had completed and described their main current or former occupation, including the activities and/or duties associated with that job. A weighted two-factor (education and occupation) individual SES score was determined, using the Hollingshead (1957) scoring procedure. Subjects were then classified at five SES levels (1 = highest level, 5 = lowest level), based on their computed two-factor score (Hollingshead & Redlich, 1958). The occupational status scale developed by Hollingshead was useful in this study given the older age of our subjects and their residence in a non-highly urbanized state.

Because the distribution of SES by subjects' race indicated that older blacks were much more likely to be at the two lowest SES levels, we defined higher SES in our final logistic regression model as levels 1–3 among blacks and as levels 1–2 among whites. Byrne, Nedelman and Luke (1994), similarly, used *race-specific* median family income of patients' zip code as their measure of the SES of black and white ESRD patients in New York State, and Young et al. (1994) used *race-specific* per capita income of patients' county of residence as a surrogate measure of SES.

Analytic Strategy

Cases and controls were compared on univariate crosstabulations using the chi-square test. Age of cases and controls was compared by a two-sample t-test. Stepwise logistic regression was used to determine whether SES was related to ESRD, after controlling for demographic characteristics and history of diabetes (SAS, 1990). Models were fit: (1) for blacks and whites separately, and (2) combining blacks and whites, including race as an additional independent variable. Two-factor interactions were assessed for potential significance, conditional on all main effects being forced into the model. All models were assessed for goodness of fit (Hosmer & Lemeshow, 1989).

RESULTS

Selection of cases and controls was designed to yield similar distributions in each group on age, sex, and race. Cases and controls did not differ on mean age (68.6 for cases, 68.0 for controls; $p = 0.25$) or on percentage male (52.5% for cases, 52.9% for controls; $p = 0.89$). However, mortality that occurred among cases in the four- to nine-month period between sample identification and subjects' interview resulted in a significant race difference between interviewed cases and controls: 58.3% of cases, compared to 47.3% of controls, were black; $p = 0.005$). Black ESRD patients have significantly better survival than do white ESRD patients on dialysis in the U.S. (USRDS, 1999).

Given the significant race difference between cases and controls, univariate associations of diabetes and SES with ESRD were investigated separately for blacks and whites. A history of diabetes was significantly associated with ESRD among both older blacks ($p < 0.0001$) and older whites ($p = 0.007$). Over half of older black female cases (52.3%) had a history of diabetes.

Black persons were heavily concentrated at the lowest SES level. In the control sample, black females were significantly more likely than males to be at a higher SES level ($p = 0.002$), while among ESRD cases, females and males did not differ significantly on SES ($p = 0.61$). Most important, when males and females were combined, black ESRD cases had significantly lower SES than did black controls ($p = 0.05$).

White persons were concentrated at the two lowest SES levels. Among white controls, males and females did not differ significantly on SES ($p = 0.73$). Among white ESRD cases, however, females were significantly lower than were males on SES ($p = 0.028$). Unlike the findings for black subjects, when

Table 2a. Race/Sex Distribution, ESRD Patients and Controls

Race/Sex	ESRD *n*	ESRD %	Controls *n*	Controls %
Black Male	102	31.3	81	24.2
Black Female	88	27.0	77	23.1
White Male	69	21.2	96	28.7
White Female	67	20.5	80	24.0
Total	326		334	

Table 2b. Mean Age (SD), by Race/Sex, ESRD Patients and Controls

Race/Sex	ESRD	Controls
Black Male	67.7 (5.6)	67.1 (6.2)
Black Female	68.2 (6.0)	68.4 (7.2)
White Male	70.1 (5.7)	68.3 (6.1)
White Female	68.7 (6.1)	68.2 (6.3)
Total	68.6 (5.9)	68.0 (6.5)

Table 3. Percentage of Cases and Controls with Diabetes, by Race and Sex

Race/Sex	Cases ($n = 326$)	Controls ($n = 334$)
Blacks*		
Male	28.4	14.8
Female	52.3	14.3
Total	39.5	14.6
Whites†		
Male	24.6	8.3
Female	23.9	17.5
Total	24.3	12.5

* $p < 0.0001$ for association of diabetes and ESRD among blacks.
† $p = 0.007$ for association of diabetes and ESRD among whites.

males and females were combined, white ESRD cases and controls had similar SES distributions ($p = 0.52$).

It is clear that the older blacks and older whites in this study had very different distributions of SES. Based on chi-square tests, race and SES were significantly related among cases ($p < 0.0001$) and among controls ($p < 0.0001$). These univariate descriptive results show the need to control on race, as well as on diabetes, when assessing the relation of SES to the probability of having ESRD.

Logistic regression analyses in race-specific models showed a trend toward higher SES being protective against ESRD among both blacks and whites. Estimated regression coefficients in the two models were quite similar. In the model for black subjects, the regression coefficient of –0.74 for higher SES approached statistical significance ($p = 0.11$). Similarly, in the model for white subjects, the regression coefficient of –0.69 approached statistical significance ($p = 0.13$). Diabetes and increasing age were significantly associated with ESRD in both of these race-specific models.

A final model including blacks and whites was then fit. In this combined model, higher SES was a statistically significant protective factor against ESRD ($p = 0.02$), with an estimated odds ratio of 0.48 and a 95% confidence interval of (0.25, 0.90). The estimated odds ratio for higher SES in the combined model was consistent with the estimated odds ratio for higher SES in the race-specific models, but the confidence interval was narrower in the combined model because of increased sample size. As in the race-specific models, diabetes and older age were significantly associated with ESRD, and black race was an additional main effect in the combined model. The main effects model had a very good fit as indicated by the Hosmer-Lemeshow test ($p = 0.69$). No interactions substantially improved the fit of the model.

DISCUSSION

We found that lower SES, as measured by individuals' education and occupation, was a risk factor for treated ESRD in older persons, even after controlling for history of diabetes. Income was not a useful SES measure in our study because most subjects were living on retirement or disability income. Education alone as a measure of SES is not sensitive to change that may occur after young adulthood. We chose to use information about subjects' current or former occupation as well as their completed educational level to measure SES.

The lack of substantial variability in SES that existed among both older black and older white subjects in this study increased the difficulty of

Table 4a. Distribution of SES by Sex and by ESRD Status – Blacks Only

	ESRD		Controls	
SES	Male *n* (%)	Female *n* (%)	Male *n* (%)	Female *n* (%)
1 or 2	2 (2.0)	1 (1.1)	3 (3.7)	7 (9.1)
3	3 (2.9)	2 (2.3)	1 (1.2)	6 (7.8)
4	17 (16.7)	11 (12.5)	15 (18.5)	3 (3.9)
5	80 (78.4)	74 (84.1)	62 (76.5)	61 (79.2)
Total	102	88	81	77

Table 4b. Distribution of SES by Sex and by ESRD Status – Whites Only

	ESRD		Controls	
SES	Male *n* (%)	Female *n* (%)	Male *n* (%)	Female *n* (%)
1 or 2	5 (7.5)	3 (4.5)	12 (12.5)	6 (7.5)
3	19 (27.5)	9 (13.4)	20 (20.8)	19 (23.8)
4	28 (40.6)	26 (38.8)	36 (37.5)	30 (37.5)
5	17 (24.6)	29 (43.3)	28 (29.2)	25 (31.2)
Total	69	67	96	80

Table 5. Final Logistic Regression Model Predicting ESRD Risk

Variable	Regression Coefficient (S.E.)	*p*-value	Odds Ratio	95% CI†
Intercept	–4.16 (1.15)	0.0003		
Black	0.41 (0.16)	0.012	1.51	(1.10, 2.09)
High SES*	–0.74 (0.32)	0.022	0.48	(0.25, 0.90)
Diabetes	1.08 (0.20)	<0.0001	2.94	(1.97, 4.39)
Male	0.06 (0.16)	0.727	1.06	(0.77, 1.46)
Age	–0.27 (0.09)	0.005		
Goodness of Fit (Hosmer-Lemeshow)	$p = 0.694$			

*SES = 1 or 2 for Whites and SES = 1, 2 or 3 for Blacks
† CI = Confidence Interval

identifying an effect of SES on risk of ESRD. Only 9% of the older ESRD cases and 13% of controls had more than a high school education. A semi-skilled or unskilled occupation was reported by 61% of our cases and 54% of controls. The concentration of both blacks and whites at lower SES levels reflected not only subjects' older age but also the fact that many lived in rural areas of a southern state.

National data show that black race, older age, and diabetes are associated with markedly elevated risks for developing ESRD. Males have a slightly elevated risk compared to females (USRDS, 1999). Our findings are consistent with these patterns, although the magnitude of the odds ratios for race, sex, or diabetes in our final model are based on prevalent patients and therefore should not be compared to the relative risks associated with these factors that are reported for incident patients.

Possible mechanisms for an increased risk of ESRD among the economically disadvantaged include higher exposure to occupational nephrotoxins, such as heavy metals and solvents; a higher intake of over-the-counter analgesics such as headache powders (popular in the South); overcrowded housing with increased exposure to streptococcal infections; substandard housing with greater exposure to lead paint and to *Hantavirus* organisms; and greater likelihood of high-risk behaviors such as street drug use (Perneger et al., 1995). All of these factors are potential sources of renal damage, but the list emphasizes risks associated with individuals' environment rather than their behavior.

Stress, social ties, and attitudinal orientations may also be links between social stratification and health outcomes (Williams, 1990). Hostility, depression, and social isolation may cluster in lower SES groups. Hostility may generate increased sympathetic nervous system activity and decreased parasympathetic nervous system activity, which are biologically plausible contributors to pathogenic processes (Williams, 1998). Hostility that generates stress may increase the risk of hypertension, which in turn is a strong risk factor for diabetes, cardiovascular morbidity, and renal disease (Klag et al., 1996; Luke, 1992).

A very important potential pathway linking lower SES, race and risk of ESRD is that persons of lower SES and blacks may be less able to control the ESRD clinical risk factors of diabetes and hypertension. It has been shown that the odds for developing ESRD due to diabetes are much higher for persons with incomes of less than $20,000 a year (odds ratio = approximately 4.0) compared to persons with an income of $20,000-$39,999 (odds ratio = 1.8) or an income of > $40,000 (odds ratio = 1.0) (Perneger et al., 1995). Almost all incident cases of diabetic ESRD after age 50 in both black and white persons are caused by non-insulin dependent diabetes (NIDDM), a health condition with increasing prevalence in the U.S. Although blacks appear to be at no higher risk for developing ESRD related to insulin-dependent diabetes (IDDM), there is an excess incidence of ESRD related to NIDDM among blacks compared to whites (Brancati, Whittle, Whelton, Seidler & Klag, 1992; Cowie et al., 1989). Black female ESRD cases in our study were especially likely to have diabetes (Table 3).

Black race remains a strong predictor of ESRD even after adjustment for other known risk factors, including diabetes (Brancati et al., 1992; Whittle, Whelton, Seidler & Klag, 1991). Genetic, biologic, psychologic, social, and environmental factors, as well as their interactions, may contribute to racial disparities in health outcomes such as ESRD. The finding that dark skin color is associated with elevated blood pressure only among persons at lower socioeconomic strata appears to illustrate interrelation of genetic and social/environmental factors (Klag, Whelton, Coresh, Grim & Kuller, 1991).

As indicated by our data, and as demonstrated by Mutchler and Burr (1991), an association among race, SES, and health is apparent even among subjects selected for longevity. The high rate of treated ESRD among older persons in the U.S., and the relation of SES and race to treated ESRD among older persons, are illustrations of the increasing concentration of health problems in our society among persons of lower socioeconomic status in middle and old age. Persons at higher socioeconomic levels seem to be better able to postpone significant increases in morbidity until relatively late in the life course. The challenge is to identify strategies by which the relation of age to health among persons at lower socioeconomic strata can become more similar to the relation of age to health among persons at higher socioeconomic strata (House et al., 1994).

Increased understanding of the links between race, SES, and disease requires research in which genetic, biologic, psychologic, social, and environmental factors, along with their interactions, are measured in relation to health outcomes. The high incidence of ESRD among older persons indicates the need to design this research so that cohort changes in health, resources, access to care, and service utilization can be monitored over extended time periods. A useful extension of our study would focus on investigating these factors in relation to race, SES, and incidence of ESRD within the framework of a life course perspective.

ACKNOWLEDGMENTS

Based on work supported by grant R01 AG05909 and grant R01 DK42949 from the National Institutes of Health. A preliminary report was presented at the 1995 annual meeting of the American Society of Nephrology, San Diego.

REFERENCES

Ayanian, J. Z., Cleary, P. D., Weissman, J. S., & Epstein, A. M. (1999). The Effect of Patients' Preferences on Racial Differences in Access to Renal Transplantation. *New England Journal of Medicine, 341*, 1661–1669.

Blane, D. (1995). Social Determinants of Health—Socioeconomic Status, Social Class and Ethnicity. *American Journal of Public Health, 85*, 903–905.

Brancati, F. L., Whittle, J. C., Whelton, P. K., Seidler, A. J., & Klag, M. J. (1992). The Excess Incidence of Diabetic End-Stage Renal Disease Among Blacks: A Population-Based Study of Potential Explanatory Factors. *Journal of the American Medical Association, 268*, 3079–3084.

Byrne, C., Nedelman, J., & Luke, R. G. (1994). Race, Socioeconomic Status, and the Development of End-Stage Renal Disease. *American Journal of Kidney Diseases, 23*, 16–22.

Cowie, C. C., Port, F. K., Wolfe, R. A., Savage, P. J., Moll, P. P., & Hawthorne, V. M. (1989). Disparities in Incidence of Diabetic End-Stage Renal Disease According to Race and Type of Diabetes. *New England Journal of Medicine, 321*, 1074–1079.

Evans, R. W., Manninen, D. L., Garrison, L. P. Jr., Hart, L. G., Blagg, C. R., Gutman, R. A., Hull, A. R., & Lowrie, E. G. (1985). The Quality of Life of Patients with End-Stage Renal Disease. *New England Journal of Medicine, 312*, 553–559.

Freeman, H. P., & Payne, R. (2000). Racial Injustice in Health Care. *New England Journal of Medicine, 342*, 1045–1047.

Hollingshead, A. B. (1957). *Two Factor Index of Social Position*. New Haven, CT: Privately printed.

Hollingshead, A. B., & Redlich, F. C. (1958). *Social Class and Mental Illness*. New York: Wiley, pp.398–407.

Hosmer, D. W., & Lemeshow, S. (1989). *Applied Logistic Regression*. New York: Wiley.

House, J. S., Lepkowski, J. M., Kinney, A. M., Mero, R. P., Kessler, R. C., & Herzog, A. R. (1994). The Social Stratification of Aging and Health. *Journal of Health and Social Behavior, 35*, 213–234.

Klag, M. J., Whelton, P. K., Randall, B. L., Neaton, J. D., Brancati, F. L., Ford, C. E., Shulman, N. B., & Stamler, J. (1996). Blood Pressure and End-Stage Renal Disease in Men. *New England Journal of Medicine, 334*, 13–18.

Klag, M. J., Whelton, P. K., Randall, B. L., Neaton, J. D., Brancati, F. L., & Stamler, J. (1997). End-Stage Renal Disease in African-American and White Men: 16-Year MRFIT Findings. *Journal of the American Medical Association, 277*, 1293–1298.

Klag, M. J., Whelton, P. K., Coresh, J., Grim, C. E., & Kuller, L. H. (1991). The Association of Skin Color with Blood Pressure in US African Americans with Low Socioeconomic Status. *Journal of the American Medical Association, 265*, 599–602.

Krieger, N. (1991). Women and Social Class: A Methodological Study Comparing Individual, Household, and Census-based Measures as Predictors of Black/White Differences in Reproductive History. *Journal of Epidemiology and Community Health, 45*, 35–42.

Krieger, N. (1992). Overcoming the Absence of Socioeconomic Data in Medical Records: Validation and Application of a Census-Based Methodology. *American Journal of Public Health, 92*, 703–710.

Krieger, N., Rowley, D. L., Herman, A. A., Avery, B., & Phillips, M. T. (1993). Racism, Sexism, and Social Class: Implications for Studies of Health, Disease, and Well-being. *American Journal of Preventive Medicine, 9* (Suppl 2), 82–122.

Krieger, N., Van Den Eeden, S. K., Zava, D., & Okamoto, A. (1997). Race/Ethnicity, Social Class, and Prevalence of Breast Cancer Prognostic Biomarkers: A Study of White, Black, and Asian Women in the San Francisco Bay Area. *Ethnicity & Disease, 7*, 137–149.

Kutner, N. G. (1994). Is There Bias in the Allocation of Treatment for Renal Failure? In: H. M. McGee & C. Bradley (Eds), *Quality of Life Following Renal Failure: Psychosocial Challenges Accompanying High Technology Medicine* (pp. 153–168). London: Harwood Academic Publishers.

Lantz, P. M., House, J. S., Lepkowski, J. M., Williams, D. R., Mero, R. P., & Chen, J. (1998). Socioeconomic Factors, Health Behaviors, and Mortality: Results from a Nationally Representative Prospective Study of U.S. Adults. *Journal of the American Medical Association, 279*, 1703–1708.

Liao, Y., McGee, D. L., Kaufman, J. S., Cao, G., & Cooper, R. S. (1999). Socioeconomic Status and Morbidity in the Last Years of Life. *American Journal of Public Health, 89*, 569–572.

Link, B. G., & Phelan, J. (1995). Social Conditions as Fundamental Causes of Disease. *Journal of Health and Social Behavior*, Extra Issue, 80–94.

Luke, R. G. (1992). Can We Prevent End-Stage Renal Disease Due to Hypertension or to Diabetes Mellitus? *Journal of the American Medical Association, 268*, 3119–3120.

Mutchler, J. E., & Burr, J. A. (1991). Racial Differences in Health and Health Care Utilization in Later Life: The Effect of Socioeconomic Status. *Journal of Health and Social Behavior, 32*, 342–356.

Navarro, V. (1990). Race or Class Versus Race and Class: Mortality Differentials in the United States. *Lancet, 336*, 1238–1240.

Pappas, G. (1994). Elucidating the Relationships Between Race, Socioeconomic Status, and Health. *American Journal of Public Health, 84*, 892–893.

Perneger, T. V., Whelton, P. K., & Klag, M. J. (1995). Race and End-Stage Renal Disease: Socioeconomic Status and Access to Health Care as Mediating Factors. *Archives of Internal Medicine, 155*, 1201–1208.

SAS/STAT User's Guide (1990). Version 6, Fourth Edition, Volume 2. Cary, NC: SAS Institute.

Shulman, N. B., Martinez, B., Brogan, D., Carr, A. A., & Miles, C. G. (1986). Financial Costs as an Obstacle to Hypertensive Therapy. *American Journal of Public Health, 76*, 1105–1108.

United States Renal Data System (USRDS) (1999). *USRDS 1999 Annual Data Report.* Bethesda, MD: National Institute of Diabetes and Digestive and Kidney Diseases, The National Institutes of Health.

Whittle, J. C., Whelton, P. K., Seidler, A. J., & Klag, M. J. (1991). Does Racial Variation in Risk Factors Explain Black-White Differences in the Incidence of Hypertensive End-Stage Renal Disease? *Archives of Internal Medicine, 151*, 1359–1364.

Williams, D. R. (1990). Socioeconomic Differentials in Health: A Review and Redirection. *Social Psychology Quarterly, 53*, 81–99.

Williams, R. B. (1998). Lower Socioeconomic Status and Increased Mortality: Early Childhood Roots and the Potential for Successful Interventions. *Journal of the American Medical Association, 279*, 1745–1746.

Wolfe, R. A., Ashby, V. B., Milford, E. L., Ojo, A. O., Ettenger, R. E., Agodoa, L. Y. C., Held, P. J., & Port, F. K. (1999). Comparison of Mortality in All Patients on Dialysis, Patients on Dialysis Awaiting Transplantation, and Recipients of a First Cadaveric Transplant. *New England Journal of Medicine, 341*, 1725–1730.

Young, E. W., Mauger, E. A., Jiang, K-H., Port, F. K., & Wolfe, R. A. (1994). Socioeconomic Status and End-stage Renal Disease in the United States. *Kidney International, 45*, 907–911.

HIV/AIDS PATIENTS' PREFERENCES REGARDING PHYSICIAN GENDER, SEXUAL ORIENTATION, AND RACE/ ETHNICITY

Kathleen Johnston Roberts

ABSTRACT

This study explored HIV-positive patients' preferences and beliefs about physician gender, sexual orientation, and race/ethnicity. In-depth interviews were conducted with 28 HIV-positive patients, including 15 men and 13 women. Results showed that over half (54%) of respondents had a preference regarding physician gender. Of these respondents, 60% (six women and three men) preferred to have women as their providers. Forty-three percent of respondents had a preference regarding physician sexual orientation. Of these eight men and four women, most (75%) stated that they preferred to see a homosexual health care professional as compared to a heterosexual one. Fewer respondents (18%) had a preference regarding physician race/ethnicity; however, all of those who had a preference regarding physician race/ethnicity stated that they preferred to see Caucasian physicians. Results suggest that physician gender, sexual orientation, and race/ethnicity are not neutral characteristics in the health care setting. Rather, these are dynamic variables that may have important

Health, Illness, and Use of Care: The Impact of Social Factors, Volume 18, pages 167–179.

ISBN: 0-7623-0740-4

implications for physician-patient relationships in HIV/AIDS care. Patients are most apt to receive optimal health care when they are able to find a physician with the sociodemographic characteristics that they prefer.

INTRODUCTION

Historically, most physicians in the United States have been white men. Yet, changes have occurred during the past few decades that have altered the sociodemographic profile of the physician work force. Perhaps the most dramatic of these relates to physician gender. For example, between 1980 and 1994, the total number of women physicians grew by 145.5% (Randolph, Seidman & Pasko, 1995). Women currently account for 20% of the entire physician work force (Randolph et al., 1995). While efforts also have been made to increase the number of racial/ethnic minorities in medicine, these efforts have resulted in less significant changes. For example, only 3.7% of all physicians in the United States are African-American and 5.5% are Hispanic (Lewin & Rice, 1994). Data regarding physician sexual orientation have never been collected; however, general population estimates suggest that approximately 10% of the general population is homosexual (Kinsey, Pomeroy & Martin, 1948). If this proportion holds true for physicians, then probably 10% of all physicians are homosexual. These percentages illustrate that the current health care environment is composed of a diverse range of physicians in terms of gender, race/ethnicity, and sexual orientation.

While a great deal of research suggests that patient sociodemographic characteristics affect physician-patient relationships (Waitzkin, 1991; Fisher, 1989; Krieger, Rowley, Herman, Avery & Phillips, 1993; Roter & Hall, 1992), less is known about the effects of physician characteristics on these relationships. Theoretical literature regarding physician-patient relationships rarely even addresses physician gender, sexual orientation, or race/ethnicity (Parsons, 1951; Freidson, 1970; Hayes-Bautista, 1976; Szasz & Hollender, 1956) and few empirical studies exist which investigate how these characteristics may affect physician-patient relationships.

Of all physician sociodemographic characteristics, gender is perhaps the one that researchers have examined the most. Studies show that women physicians spend more time than men physicians with patients (Cypress, 1980) and that much of the excess time is spent talking with patients about their medical histories (Roter, Lipkin & Korsgard, 1991). Not surprisingly, research also shows that patients of female physicians are more satisfied than patients of male physicians with their health care (Delgado, Lopez-Fernandez & Luna, 1993). This higher level of patient satisfaction may contribute to the fact that women physicians are less likely than men physicians to be sued for malpractice (Taragin, Wilczek, Karns, Trout & Carson, 1992).

Do patients have preferences regarding physicians' gender? Nichols (1987) and Graffy (1990) found that women often prefer to see female physicians for sex-related problems, gynecological examinations, and emotional problems. Elstad (1994) found that women patients who value forming egalitarian relationships with their physicians prefer to see female doctors.

Some research also suggests that some male patients have preferences regarding physician gender. For example, Heaton and Marquez (1990) surveyed 72 male patients about their preference to have either a man or woman physician perform genital and rectal examinations on them. Results showed that 52% of subjects preferred a male physician for the genital exam, while 48% indicated no preference. For the rectal exam, 60% had no preference for physician gender while 40% preferred a male physician. No subjects preferred a female physician for either exam.

Less research has been conducted regarding patients' preferences regarding physician sexual orientation and race/ethnicity, but some evidence suggests that these too may be important variables to consider. For example, Stevens (1994) interviewed an ethnically diverse sample of lesbian women about their health care experiences. She found that having a physician who was "a mirror of themselves" in terms of sociodemographic characteristics made them feel more secure in health care settings. One participant in her study stated, "I found a Black lesbian doctor, can you believe it? She can relate to me." (Stevens, 1994: 222). This suggests that physician sociodemographic characteristics may have important effects on physician-patient relationships, including patients' satisfaction with health care.

The purpose of this study was to further explore patients' preferences and beliefs about physician gender, sexual orientation, and race/ethnicity. Because data were collected as part of a larger study regarding physician-patient relationships in HIV/AIDS care, HIV-positive patients served as the respondents. Because HIV/AIDS is a long-term, chronic disease where much of what medicine has to offer comes in the way of "care" vs. "cure," physician-patient relationships are especially important in this domain. Any clues regarding HIV-positive patients' preferences and beliefs regarding physician characteristics may lead to an improved overall understanding of physician-patient relationships in HIV/AIDS care.

METHODS

In-depth, face-to-face interviews were conducted with 28 HIV-positive patients. Interviews lasted approximately 60 minutes and were audiotaped. All patients were asked about their relationships with their primary care physicians and what preferences, if any, they had regarding physician gender, sexual orientation, and race/ethnicity. This chapter will report on data pertaining only to patients' preferences regarding physician sociodemographic characteristics. The interview questions were open-ended and the content, as well as the flow of conversation, changed with each subject, to match what the interviewee knew and felt (Rubin & Rubin, 1995).

Written informed consent was obtained prior to the start of each interview. Each participant was paid $20.00 in cash for his/her time and for the transportation costs associated with traveling to the interview site. Fieldnotes were written at the conclusion of each interview and all tapes were transcribed.

Sample – Recruitment and Composition

Participants were recruited from the University of California, San Francisco AIDS Program of San Francisco General Hospital (SFGH). This well-regarded, urban, public health clinic serves primarily low-income individuals from diverse racial/ethnic backgrounds. Recruitment occurred between June 1997 and January 1998. Participants were recruited through flyers handed out and posted on bulletin boards in the waiting room. To be eligible for participation, patients had to be 18 years of age, be able to speak English, and have a primary care physician. Also, only persons who self-reported that they were HIV-positive and had been taking an antiretroviral cocktail regimen that included a protease inhibitor for at least three months were eligible to participate.

Fifteen men and 13 women respondents participated. The average age of respondents was 40, although ages ranged from 19 to 54. About half (46%) of respondents were Caucasian, 36% were African-American, 14% were Hispanic and 4% were of other racial/ethnic backgrounds. Over half (54%) of respondents had completed at least some college education while the rest had completed high school or less education. Forty-three percent of respondents earned less than $5,000.00 per year, 39% earned between $5,001.00 and $10,000.00 per year, 14% earned more than $10,0001.00 per year, and 4% did not know their annual income. The majority (68%) of patients had MediCal insurance, 18% had both MediCal and MediCare, 4% had private health insurance, and 11% had no insurance.

The year of HIV diagnosis for patients ranged from 1984 to 1996; the modal year of diagnosis was 1989 ($N = 5$). Eleven-percent of patients rated their health status as "excellent," 54% as "good," 32% as "fair," and 4% as "poor." The majority (75%) of respondents had HIV viral loads under 5,000 copies while 8% had counts between 5,0001 and 50,000 and only one subject had a count above 100,000. Three patients did not know their viral loads. Eighteen-percent of patients had less than 100 T-cells, 43% had 101–250, 32% had 251–500, and 8% had 501 or more T-cells. The lower an HIV-positive patient's HIV viral load and the higher his/her T-cell count, the healthier he/she is. High viral loads and low T-cell counts are correlated with morbidity and disease progression in HIV/AIDS patients.

Analysis

Data were analyzed using standard qualitative techniques, including contact summary sheets, inductive generation of coding categories, marginal remarks and memos (Strauss & Corbin, 1990; Miles & Huberman, 1994). Each of these techniques will be described in more detail. According to Miles and Huberman (1994: 51), "A contact sheet is a single sheet with some focussing or summarizing questions about a field contact." In this study, after each interview, a contact sheet was written which answered the following questions: (1) What people were involved? (2) What was the "feel" of the interview (i.e. rapport, unusual circumstances, unexpected happenings, etc.)? (3) Which areas of inquiry in the initial study framework did the interview bear on most centrally? (4) What new hypotheses, speculations, or hunches about the study were suggested by the interview? (5) What changes, if any, should be made in subsequent interviews?

Next, the inductive generation of coding categories is called "open coding" (Strauss & Corbin, 1990). In open coding, investigators break down, conceptualize, and categorize the data. Concepts are created when investigators place a label on "happenings, events, or other instances or phenomena." (Strauss & Corbin, 1990: 61). In this project, all transcripts were perused on a line-by-line basis and codes were developed to label key themes in the data. Ethnograph (a software program for computer-based text search and retrieval) was used to help manage the data during this coding process. Specifically, all the coded segments of data were entered into Ethnograph and the program was then used to print all the data corresponding to each individual coding category.

These data were again reviewed on a line-by-line basis and sub-codes were developed for each category.

Next, marginal remarks are reflective comments that occur to a researcher as coding proceeds (Miles & Huberman, 1994). In this study, remarks were noted in the margins of the transcripts and were often further developed into memos during the analysis process. According to Strauss and Corbin (1992: 198), "memos represent the written forms of our abstract thinking about data." Memos are used for analytical purposes – they do not simply summarize data, they make sense of it. In this study, memos were used throughout the duration of the project to help with all phases of the qualitative data collection and analysis.

Adequacy and Plausibility of Data

Informal respondent validation and peer review were employed to ensure the adequacy and plausibility of the data (Lincoln & Guba, 1985; Miles & Huberman, 1994). First, throughout the interviews, the interviewer asked subjects if she was understanding what they were saying, thus providing respondents with the opportunity to correct any misunderstandings that may have arisen, and adding additional data to the project that were used to enrich the analyses and bolster the credibility of the findings. Also, the interviewer often "checked" information gleaned from one respondent with another (e.g. "Some people have said that they prefer to see a woman physician. What do you think about this?"). Next, peer review was accomplished by having the method of analysis and results critiqued by two independent medical sociologists and one physician who cared for HIV-positive patients. None of these individuals expressed concerns about the accuracy of the findings.

RESULTS

Gender

Over half (54%) of respondents reported having a preference regarding physician gender. Of these respondents, 60% (six women and three men) preferred to have women as their providers. Reasons for preferring women related to respondents' beliefs regarding physicians' superior interpersonal skills, including both affective and communicative relations with patients. To describe women physicians, respondents used adjectives such as "caring," "compassionate," "sensitive," and "nurturing;" these adjectives were not typically used to describe men physicians.

Several respondents remarked on a certain "maternal instinct" that many women physicians have. A 28-year-old Caucasian man stated, "She's more patient than most of the male doctors that I've seen. . . . She's very stern, but in a loving way . . . very much like a mother." In a similar fashion, a 35-year-old Caucasian man maintained:

> I think sometimes I like women's approach a little better than men's. . . . I mean . . . I guess, usually my experience with going to the doctor is [laughter] I really feel like shit and I need to feel better, you know, so I mean maybe it's like a "mother thing" or something. . . . Usually they're a bit more conversational.

Some of the women respondents remarked that they preferred to see female physicians because they believed that such individuals were more knowledgeable, aware, and concerned about female-specific health problems than were male physicians. Summarizing this point of view, a 37-year-old African-American woman stated:

> I feel more comfortable with a woman. . . . I just feel like the men are used to, or treat, male patients differently. HIV still is not a woman's disease, although women are the most being infected, at this point.

Women respondents also mentioned that they felt safer and more secure undressing in front of female health care professionals and that discussing personal matters, such as those related to sexual health and well-being, was also easier with women providers. For example, a 49-year-old Caucasian woman disclosed:

> I've mostly had men, just because when I was younger there wasn't many women doctors. . . . I wouldn't tell them if . . . I was bleeding too much or something. I wouldn't tell them because I was too embarrassed. And then when they want to do breast exams, I always tell them, "Well, I'll do it myself." So – but a woman, I didn't mind her doing it.

Several of the women respondents reported that they preferred to see women physicians because of a personal or family history of abuse inflicted by men. For example, a 34-year-old African-American woman stated:

> You can relate better with a female . . . your own sex. . . . I found out having a male doctor, I kind of like distance myself from them and I guess it has something to do with problems that I've gone through, you know, my childhood and everything . . . things that happened to me by men or a man or whatever.

No women respondents stated that they preferred to see male physicians. In contrast, six men stated that they preferred to see male physicians. The basic theme was that male physicians could relate better to male patients because both shared a similar "mindset." An excerpt from a 47-year-old Caucasian man's interview provides illustration:

Interviewer (I): Some people say that they prefer to see either a man or a woman physician.
Respondent (R): Absolutely a man.
I: A man. Tell me why.
R: Because he's the same sex and gender, and mindset of a male.
I: What does that mean?
R: Security, safe.
I: It's safer to see a man?
R: Just feels like they'll hear me.
I: In a way that's different than [the way] a woman would?
R: Yeah, because . . . men are from Mars and women are from Venus.

In addition to this "mindset" issue, one male patient said he preferred to see a male physician because he felt more uncomfortable undressing in front of such a provider as compared to a woman.

Sexual Orientation

Forty-three percent of respondents reported having a preference regarding physician sexual orientation. Of these eight men and four women, most (75%) stated that they preferred to see a homosexual health care professional as compared to a heterosexual one. Although data regarding respondents' sexual orientation were not collected as part of this project, many of the respondents who reported preferring a gay physician self-identified as being homosexual or bisexual.

Many of those who preferred gay physicians stated that such providers would be more sympathetic to their needs. For example, a 19-year-old African-American man stated, "I feel much better with a gay. . . . physician. . .because I'm gay. . .they know exactly what I'm going through." Many of these respondents also reported that they would feel more comfortable discussing personal issues with these providers. As a 37-year-old African-American woman stated:

R: She's a lesbian. . . . I can talk to her about my relationship, I'm dating this woman and yada, yada, yada, "What do you think about safe sex? And what do you think – how about if I did this? Would that be unsafe?"
I: Would you say those things with a straight physician?
R: Oh, no. . . . No. Because they wouldn't know. What can they tell me? Straight physicians, they have no idea what two women even do. How can they give you advice?

Not all respondents reported preferring a homosexual physician. One man and two women reported that they would prefer to see a heterosexual doctor. The man, a 35-year-old Hispanic individual, had had a negative experience in the past with a physician whom he had presumed to be gay. As he stated, "I think

I would prefer to see a straight physician. . . . I've had an experience in the past with a physician who seemed to just enjoy my company, I think more than he should have." Both women who preferred to see a heterosexual physician admitted to feeling uncomfortable around gay people. As one, a 34-year-old African-American individual, stated, "I guess I've always felt a bit uncomfortable around the gays. I really don't know how to act and what to talk about. . . . I've never really been around them growing up or anything."

Race/Ethnicity

Compared to physician gender and sexual orientation, fewer respondents (18%) reported having a preference regarding physician race/ethnicity. When asked about this topic, many patients responded by saying such things as "I don't discriminate." Some respondents noted that it was a difficult question to answer as they had little experience with non-white physicians. For example, a 53-year-old Caucasian man noted, "I've always had white doctors. . . . I've never come across a situation where I had to go to another ethnically-oriented doctor."

All of those who had a preference regarding physician race/ethnicity stated that they preferred to see Caucasian physicians. These patients, four of whom were Caucasian and one of whom was African-American, remarked that they preferred such physicians simply because that was what they were used to. As a 47-year-old Caucasian man, explained, "I prefer a Caucasian. . . . It's what I'm used to. It's, you know, that I can communicate with them better, and feel better with them."

DISCUSSION

Many respondents in this study had strong preferences for and beliefs about physician sociodemographic characteristics. These results suggest that physician gender, sexual orientation, and race/ethnicity are not neutral characteristics in the health care setting. Rather, these are dynamic variables that may have important implications for physician-patient relationships in HIV/AIDS care, including patients' willingness to disclose information during office visits.

Many respondents who reported a preference regarding physician gender preferred to see women physicians. This finding is not surprising given previous research which shows that women physicians spend more time with patients and express a greater interest in the interpersonal aspects of patient care as compared to men physicians (Cypress, 1980; Roter et al., 1991; West, 1993). Psychosocial aspects of patient care, such as talking with patients and expressing an interest in patients' overall well-being (rather than just to their signs and symptoms of disease), are likely of integral importance in chronic, long-term disease management such as occurs in HIV/AIDS care.

Women respondents who preferred to see a female physician believed that such persons were better equipped to handle female-specific health problems. These women also believed that talking about personal matters (e.g. sexual health and well-being) and undressing during office visits was easier with women doctors as compared to men doctors. Women who had been abused by a man in the past were particularly likely to prefer to see a woman physician. Three men respondents also expressed a preference to see a female physician, stating that they enjoyed women physicians' personal, nurturing style of communication. Other men respondents preferred to see male physicians, mainly because they believed that such persons shared a similar male "mindset" with themselves. It may be that men who prioritize a highly interpersonal style of physician-patient communication may be most satisfied with a female physician while those who simply are more comfortable with a man due to the "mindset" issue may prefer to see a male physician.

Many male respondents had strong preferences regarding physician sexual orientation. HIV-positive gay men are well known for being particularly well-informed, assertive patients (Institute of Medical Ethics, 1992); such assertiveness may make these men especially attentive to the interpersonal aspects of physician-patient interactions, including what physician characteristics increase their own comfort levels. A couple lesbian respondents also reported preferences regarding physician sexual orientation. This finding coincides with research which suggests that many lesbian and bisexual women prefer to see lesbian health care professionals (Reagan, 1981; Paroski, 1987; Stevens, 1994).

Previous studies have documented the difficulties (e.g. feeling uncomfortable, being intimidated by health care professions, hesitating to disclose information, etc.) that many homosexual individuals endure in the health care system (Stevens, 1992; Cochran & Mays, 1988; Trippet, 1992). If having a homosexual physician improves gay and lesbian patients' health care experiences, efforts should be made to refer patients to such providers whenever possible and to increase the proportion of heterosexual health care professionals who can provide compassionate, non-biased, and respectful care to gay and lesbian patients.

Although fewer respondents expressed preferences and beliefs regarding physician race/ethnicity than sexual orientation or gender, these findings may have been different if non-English speaking respondents had been included in the sample. Such individuals may prefer to see physicians from their own racial/ethnic groups who speak their native languages fluently. Future research should investigate non-English speaking individuals' preferences and beliefs regarding physician race/ethnicity.

The fact that many respondents did not report a preference regarding physician race/ethnicity and made statements such as "I don't discriminate" highlight the often taboo nature of the topic of race in the United States. Events such as the Rodney King beating and the O. J. Simpson trial have intensified many individuals' uneasiness about discussing this area. The taboo nature of the topic may have adversely affected respondents' willingness to disclose their opinions. Social desirability biases also may have led some to say they did not have a preference regarding physician race/ethnicity when perhaps they did. Conversely, it may be that respondents simply had not been exposed to many non-Caucasian physicians in the health care system and therefore had not developed strong preferences and beliefs about physician race/ethnicity.

While many respondents were interested in finding a "mirror of themselves" (Stevens, 1994) in their health care professionals, some respondents specifically did not want their physician to be like them in terms of gender, sexual orientation, or race/ethnicity. Future research should investigate the differences between patients who do and do not prefer to see a physician with similar sociodemographic characteristics to themselves. Also, patients' preferences regarding physician gender, sexual orientation, and race/ethnicity must be distinguished from their preferences regarding other variables, such as physician communication style.

While findings from this study provide a rich illustration of some patients' preferences regarding physician sociodemographic characteristics, some limitations to the study should be mentioned. First, the sample for this project was small and the respondents consisted entirely of HIV/AIDS patients. Future studies should attempt to replicate these findings with larger samples and/or other types of patients, such as those living with cancer, or those seeking routine/preventive health care. Next, the study was conducted in San Francisco, a unique, urban environment on the West Coast of the United States. Individuals in other geographical locales may have different preferences about physician gender, sexual orientation, and race/ethnicity.

Finally, it is imperative to note that patients' preferences about physician sociodemographic characteristics may not necessarily be consistent with their actual use of a physician. Some patients may prefer to see a woman physician, but may be unable to do so because the clinic they attend has only men physicians on staff. Other patients may be limited in their choice of physician by their health insurance plan. Because racial/ethnic minorities are under-represented in medicine (Lewin et al., 1994), some patients may have difficulty locating a physician of a specific race/ethnicity. Nevertheless, findings from this study suggest that patients are most apt to receive optimal health care when they are able to find a physician with the sociodemographic characteristics that they prefer.

ACKNOWLEDGMENTS

This research was supported by funds from the Universitywide AIDS Research Program, University of California, grant No. D97-SF-001.

REFERENCES

Cochran, S. D., & Mays, V. M. (1988). Disclosure of sexual preference to physicians by black lesbian and bisexual women. *Western Journal of Medicine*, *149*, 616–619.

Cypress, B. K. (1980). Characteristics of visits to male and female physicians. *Vital and Health Statistics*. (Series 13, No.49). Hyattsville, MD: U.S. Department of Health and Human Services.

Delgado, A., Lopez-Fernandez, L. A., & Luna, J. D. (1993). Influence of the doctor's gender in the satisfaction of the users. *Medical Care*, *31*, 795–800.

Elstad, J. I. (1994). Women's priorities regarding physician behavior and their preference for a female physician. *Women and Health*, *21*, 1–19.

Fisher, S. (1989). *Intimate Adversaries: Cultural Conflict Between Doctors and Women Patients*. Philadelphia: University of Pennsylvania Press.

Freidson, E. (1970). *Professional Dominance*. New York: Atherton.

Graffy, J. (1990). Patient choice in a practice with men and women practitioners. *British Journal of Family Practice*, *40*, 13–15.

Hayes-Bautista, D. (1976). Modifying the treatment regimen: patient compliance, patient control, and medical care. *Social Science and Medicine*, *10*, 233–238.

Heaton, C. J. and Marquez, J. T. (1990). Patient preferences for physician gender in the male genital/rectal exam. *Family Practice Research Journal*, *10*, 105–115.

Institute of Medical Ethics Working Party on the Ethical Implications of AIDS (1992). AIDS and the ethics of medical care and treatment. *Quarterly Journal of Medicine*, *83*, 419–426.

Kinsey, A. C., Pomeroy, W., & Martin, C. (1948). *Sexual Behavior in the Human Male*. Philadelphia: W. B. Saunders, Co.

Krieger, N., Rowley, D. L., Herman, A. A., Avery, B., & Phillips, M. T. (1993). Racism, sexism, and social class: implications for studies of health, disease, and well-being. *American Journal of Preventive Medicine*, *9* (Suppl), 82–122.

Lewin, M. E., & Rice, B. (Eds) (1994). *Balancing the Scales of Opportunity: Ensuring Racial and Ethnic Diversity in the Health Professions*. Washington, D.C.: National Academy Press.

Lincoln, Y. S., & Guba, E. G. (1985). *Naturalistic Inquiry*. Newbury Park: Sage Publications.

Miles, M. B., & Huberman, A. M. (1994). *Qualitative Data Analysis: an Expanded Sourcebook*. Thousand Oaks: Sage Publications.

Nichols, S. (1987). Women's preferences for sex of doctor: A postal survey. *Journal of the Royal College of General Practitioners*, *37*, 540–543.

Paroski, P. A. (1987). Health care delivery and the concerns of gay and lesbian adolescents. *Journal of Adolescent Health Care*, *8*, 188–192.

Parsons, T. (1951). *The Social System*. New York: Free Press.

Randolph, L., Seidman, B., & Pasko, T. (1995). *Physician Characteristics and Distribution in the U.S.* American Medical Association.

Reagan, P. (1981). The interaction of health professionals and their lesbian clients. *Patient Counseling and Health Education, 3*, 21–25.

Roter, D. L., & Hall, J. A. (1992). *Doctors Talking with Patients/Patients Talking with Doctors: Improving Communication in Medical Visits*. Westport, CN: Auburn House.

Roter, D., Lipkin, M., & Korsgaard, A. (1991). Sex differences in patients' and physicians' communication during primary care medical visits. *Medical Care, 29*, 1083–1093.

Rubin, H. J., & Rubin, I. S. (1995). *Qualitative Interviewing: The Art of Hearing Data*. Thousand Oaks: Sage Publications.

Stevens, P. E. (1992). Lesbian health care research: a review of the literature from 1970 to 1990. *Health Care for Women International, 13*, 91–120.

Stevens, P. E. (1994). Protective strategies of lesbian clients in health care environments. *Research in Nursing and Health, 17*, 217–229.

Stewart, M., Brown, J. B., Weston, W. W., McWhinney, I. R., McWilliam, C. L., & Freeman, T. R. (1995). *Patient-Centered Medicine: Transforming the Clinical Method*. Thousand Oaks: Sage Publications.

Strauss, A., & Corbin, J. (1990). *Basics of Qualitative Research*. Newbury Park: Sage Publications.

Taragin, M. I., Wilczek, A. P., Karns, M. E., Trout, R., & Carson, J. L. (1991). Physician demographics and the risk of malpractice. *The American Journal of Medicine, 93*, 537–542

Trippet, S. E. (1992). Reasons American lesbians fail to seek traditional health care. *Health Care for Women International, 13*, 145–153.

Waitzkin, H. (1991). *The Politics of Medical Encounters: How Patients and Doctors Deal with Social Problems*. New Haven: Yale University Press.

West, C. (1993). Reconceptualizing gender in physician-patient relationships. *Social Science and Medicine, 36*, 57–66.

PART III

HEALTH CARE SYSTEM ISSUES: HEALTH POLICY, HEALTH LIFESTYLES AND THE CONSTRUCTION OF PATIENTS, ILLNESS AND DEATH

CONSTRUCTING THE "GOOD PATIENT" IN CONVENTIONAL AND UNCONVENTIONAL MEDICAL SETTINGS: ROLES, RELATIONSHIPS, AND INFORMATION TRANSFER

Marian Katz

ABSTRACT

This chapter is about the work that patients do to present themselves to their doctors as certain kinds of patients. The concept of "the task of doing: 'being a good patient'" is used to emphasize that patient self-presentation in medical interactions is not a static patient attribute but is rather a dynamic activity that patients and doctors negotiate during the medical encounter. Data segments of doctor-patient symptom-reporting interactions from a conventional and an unconventional medical setting are used to examine how differences in patient symptom reports in the two settings are linked to differences in the ways the patient and doctor orient to "the task of doing: 'being a good patient'." The analysis of these data segments suggests further that asymmetry in interaction in conventional settings limits, while mutuality in the unconventional setting promotes, the transfer of information between doctor and patient.

Health, Illness, and Use of Care: The Impact of Social Factors, Volume 18, pages 183–206.
2000 by Elsevier Science Inc.
ISBN: 0-7623-0740-4

Patients don't want to tell everything to their mainstream doctors because they feel they're being judged. For example, I tried unsuccessfully for a long time before becoming pregnant with my second child. We had even tried adoption, and when that fell through I was ready to give up altogether. Then I did become pregnant again, however several months into the pregnancy I began to experience a lot of bleeding.

I talked with my doctor and he told me, "Probably you'll lose it, but you can stay in bed if you want to try to keep it." Then he added, "And don't do anything weird." Again, I was ready to give up. Talking with my doctor made me feel really helpless. But I really wanted the baby, so I thought, "what the heck," and I called this astrologist who I knew and liked. She told me that she really didn't know what would happen, but that she could give me some advice. She told me that if I really wanted this baby that I had to embrace it and welcome it into the world.

This advice was very empowering, and so I did embrace the baby. I embraced it and I welcomed it. I also called some health food stores and found out that Vitamin K might be helpful because it coagulates the blood. So I got some Vitamin K. Gradually the bleeding stopped and in the end my son was born – a healthy baby and a normal birth. Of course I never told my doctor about the "weird" things I had done. I'll never know whether or not they saved my baby, but I'm glad that I did them.

INTRODUCTION

The above anecdote offers a glimpse into the conscious life of the patient. It illustrates vividly how an act of communication can result in a patient feeling helpless or empowered, discouraged from talking or encouraged to talk about her use of unconventional therapies. The doctor's negativity in general ("probably you'll lose it") and towards unconventional medicine in particular ("don't do anything weird") ensure that the patient will withhold information about her use of unconventional therapies from her conventional doctor ("Of course I never told my doctor about the 'weird' things I had done").

This chapter is about the work that patients do to present themselves to their doctors as certain kinds of patients. I use the phrase, "the task of doing: 'being a good patient'," to emphasize that patient self-presentation in medical interactions is not a static patient attribute but is rather a dynamic activity that patients and doctors negotiate during the medical encounter. The paper focuses on how patient reports of self-observation in conventional and unconventional medical settings are shaped by "the task of doing: being a good patient," which is in turn influenced by the mutually embedded technical and social concerns present in the encounter.

PATIENT PASSIVITY AS AN INTERACTIONAL ACCOMPLISHMENT

Many patients today, like the woman quoted above, are consumers of both conventional and unconventional medicine. Current understandings of the role of the patient in the social science and medical literatures, however, are inadequate for explaining patient contributions in doctor-patient interactions in either conventional or unconventional medicine. Even when they acknowledge patients as important participants in doctor-patient interaction, most researchers direct their attention to the power of the doctor to determine the course of the interaction (Byrne & Long, 1976; Frankel, 1990; Mishler, 1984; Roter & Hall, 1992). This decision may reflect the lesser importance placed on the contributions of patients to interactions and/or the lack of a method with which to study them. Power and powerlessness, like activity and passivity, tend to be perceived as attributes of individuals rather than as products of interaction, that is, as relational and processual phenomena. This helps to explain the overattention to the doctor's role in doctor-patient interaction, and the corresponding underattention to patient contributions.[1] Ethnomethodology provides conceptual and linguistic tools for attending to the dynamics of interactional processes and for making the taken-for-granted constituents of interaction available for observation and analysis. In this study, then, I employ an ethnomethodological sensibility to compare the ways that patients use a variety of communicative devices and strategies to "do: being good patients" in two types of medical settings and to demonstrate that the transfer and withholding of information is a central part of these processes.

The inadequate conceptualization of the patient in the social science literature is paralleled in the medical literature. In a widely cited *New England Journal of Medicine* article reporting the results of an investigation into Americans' use of unconventional therapies, for example, the authors express concern about the state of doctor patient communication. They report, "Our observation that the majority of users of unconventional therapy did not discuss this therapy with their medical doctors suggests a deficiency in current patient-doctor relations" (Eisenberg et al., 1993: 252).[2] The authors entertain only one understanding of this 'deficiency': patients fail to report their use of unconventional therapies to their medical doctors because their doctors don't ask them to. According to the authors, the reasons doctors fail to ask about their patients' use of unconventional therapies are: first, that they mistakenly assume their patients are not using unconventional therapies; and second, that they lack adequate knowledge of these therapies to ask about them. The solution, then, becomes solely a matter of continuing medical education. Improved doctor-patient communication will result from physicians acquiring additional information: about unconventional therapies; about the prevalence of these therapies; and about the importance of asking patients about their use of these therapies.

The doctor-centered perspective of the *NEJM* authors treats the transfer of information by patients as simply a reaction to doctors' actions, thus missing the point that even taking a passive role in interaction is an accomplishment requiring active and strategic participation on the part of the patient. The problem with missing this point is that it results in a distorted assessment of the action of patients. In particular, it elides the possibility that patients may reasonably and even intentionally monitor the information they provide to physicians. Such a possibility is clearly excluded by the admonishment near the end of the *NEJM* article that: ". . . this failure to communicate is not in the best interest of the patients, since the use of unconventional therapy, especially if it is totally unsupervised, may be harmful" (Eisenberg et al., 1993: 252). Leaving debates about the safety or harm of conventional medical treatments aside, the tone of this passage is undeniably patronizing. The content, moreover, suggests not only that patients are incapable of exercising good judgment about self care, but also, by omitting any mention of unconventional practitioners (I will refer to them also as doctors below), that conventional doctors should, and are the best qualified to, oversee patient use of unconventional therapies.

CONSIDERING THE PATIENT EXPERIENCE

Is the experience of patients in conventional medical settings likely to lead them to entrust their conventional doctors with oversight of their use of unconventional medicine? What do patients learn about their roles and their relationships to doctors from encounters in conventional medical settings? From the quotes above and data below we might come up with something like the following: Patients experience being judged by doctors (particularly, here, regarding their reliability as informants about their own bodies); from which they learn that they are considered incompetent to understand their own bodies and to make treatment decisions, and that they are in danger of being denied the recognition and assistance they seek if they fail to portray themselves as reliable informants and compliant help-seekers. They experience doctors employing communicative devices and strategies that create and maintain distance between patient and doctor, that encourage patients to take a passive stance toward observing their bodies, and that reward patients for relinquishing responsibility for understanding and treating their bodies to the doctor.

Granted, the actual experiences of patients with doctors in conventional settings fall along a wide-ranging continuum from highly mutual to highly asymmetrical. Still, there is ample evidence to show that the orientations described above are prevalent, and that they are encouraged by both medical training and the institutional organization of medical care. It is also known that certain styles of doctor-patient communication associated with these orientations have been linked to problems faced by conventional doctors, ranging from patient noncompliance to malpractice claims. Given, on one hand, these well-known problems in doctor-patient interaction in conventional settings, and on the other hand, the prevalence of Americans' use of unconventional medical care, it should be of great interest to know more about doctor-patient interaction, and the experience of patients in general, in unconventional settings.

There has been more research on doctor-patient interaction in conventional than in unconventional medical settings. I have discussed elsewhere (Katz, 1996) that unconventional medical settings in the U.S. create certain kinds of atmospheres explicitly in relation to conventional settings – borrowing elements that they think will lend legitimacy and choosing contrasting elements to highlight the difference of what they have to offer. I have, further, shown that there is an observable, dialectical, relationship between the elements, including interactional strategies used, of conventional and unconventional medical settings. From this perspective, it is not surprising that the patient experience in the example from the unconventional setting included below contrasts sharply with the example from the conventional setting. In the unconventional setting the patient is treated as a competent informant as well as judge of her body's condition; is encouraged to share responsibility for understanding and treating her body with the doctor; and that, overall, mutuality is emphasized and communicative strategies are used to minimize social distance between doctor and patient.

In the comparison below, I use two doctor-patient interactions, one each from a conventional and an unconventional medical setting, to illustrate how doctor and patient roles and relationships come to be constructed and understood in interaction, and the role of information transfer in these processes.

DATA AND METHODS

One of the doctor-patient interactions comes from my own ethnographic research at an unconventional medical setting[3] in Southern California. My research site was a combined shop and practice, which I refer to as the Healing

Center, providing Chinese medicine[4] by and for a primarily non-Chinese population. I observed and wrote detailed field notes, usually while the interactions were taking place, on many doctor-patient interactions in this setting over an eight-month period. I especially had access to interactions that focused on symptom reporting and diagnosis. In this work I compare the style and content of one of the doctor-patient interactions that I observed at the Healing Center with a contrasting example from one of the conventional primary care settings studied by Timothy Halkowski (forthcoming).

Halkowski (forthcoming) used conversation analytic theory and methods to analyze a set of doctor-patient, symptom-reporting, interactions videotaped at two American primary care clinics, one in the Southeast and one in the Midwest. Such a cross-methodological comparison, of course, requires an extra degree of caution. Broadly speaking, ethnographers and conversation analysts are motivated and guided by different sets of concerns and commitments (see endnote 3). Because of this, I enlisted the help of scholars working in conversation analysis, and learned more about it myself. These efforts have persuaded me, on the one hand, that the existence of certain shared concerns, some of which are discussed below, along with the limited availability of high quality recorded data, makes this chapter a worthwhile contribution. On the other hand, they have also persuaded me that further research using data from recorded doctor-patient interactions in unconventional settings would be necessary to test and elaborate the ideas presented here.

DOCTORABILITY AND INFORMATION TRANSFER

One reason why the doctor role is taken more seriously than the patient role is that institutions exist with the sole purpose of producing doctors, and the processes by which individuals learn to be doctors have been well studied (Becker et al., 1961; Light, 1980; Konner, 1987). Doctors, moreover, form a well-defined and established professional group. In contrast, the training and organization of people to be patients (other than by medical providers and insurers)[5] appears to be informal, unregulated and individualized. While not formally institutionalized, the training of individuals to be patients is nevertheless structured. For instance, Parsons (1951), in his discussion of the 'sick role,' observed that norms exist regarding the ways that patients should present themselves and their concerns to doctors. In brief, Parsons noted that in order to claim the exemptions from usual social activities and from responsibility for one's illness accorded with the sick role status, one must demonstrate certain orientations towards doctors and illness. Namely, one must orient to being sick as undesirable, to the decision to go to the doctor as accountable, and to doctors' diagnoses and prescriptions as efficacious and authoritative.

Studies of doctor-patient interaction (Gill, 1998; Halkowski, forthcoming; Heritage, forthcoming) have, in fact, found that when patients present their concerns to doctors they attend as well to demonstrating the appropriateness of their orientations to: (1) observing their bodies, (2) their reason for the visit, and (3) doctors' authority. Thus patients' information transfer decisions are seen to be made in conjunction with orienting to sick role criteria. The way that a patient accomplishes these multiple tasks is by presenting himself with a concern that is 'doctor relevant' or 'doctorable'; that is, a concern that this specific doctor can identify, advise on, and/or treat. An important goal of the initial part of the doctor-patient encounter, then, is to establish why the patient is there and whether or not the reason presented meets the criteria of doctorability.

> Establishing that they have a doctorable problem is a method of justifying the decision to visit a physician. It is a means for patients to show that they are reasonable people, which in this context means showing that they have a problem or a concern for which seeking medical assistance is a reasonable solution. The presentation of a complaint determined to be non-doctorable can deprive the patient of authoritative medical support for their claim to financial and other benefits from entering the 'sick role' (Parsons, 1951, 1975; Freidson, 1970), and "engender a vulnerability to the judgment that they were misguided in seeking medical assistance . . ." (Heritage, forthcoming)

Once a patient has decided to go to the doctor, she acquires an interest in her concern being judged doctorable. This raises the further question of how patients know or learn how to present their concerns as doctorable. It should be noted that patients do not universally orient to the need to establish the doctorability of their concerns. Heritage (forthcoming), for instance, notes that both a gunshot wound and a dog bite are treated as unquestionably doctorable by both patients and doctors. Orientations towards establishing doctorability, moreover, may vary by individual and culture as well as by type of concern. In many instances, however, patients do clearly orient to the need to establish doctorability.

Among the various resources that patients may draw on for presenting themselves and their concerns to doctors, an important one is their experience engaging in 'troubles talk' in ordinary conversation and with service providers (Jefferson, 1981, 1988). One main goal of troubles telling in ordinary conversation is the establishment of emotional reciprocity (Jefferson, 1988). Jefferson (1981) notes, however, that advice-giving as a response to troubles telling in ordinary conversation is problematic because there is an inherent tension between the processes of advice-giving and of establishing emotional reciprocity. The aim of 'troubles talk' in ordinary conversation is often not problem solving but establishing an identity and relationship between the troubles teller and the troubles recipient. When advice is sought, as in a service encounter, troubles tellers refrain from talk that moves towards emotional reciprocity. They thereby exhibit troubles resistance as a way of maintaining the social distance characteristic of relationships between laypeople and professionals – i.e. advice seekers and advice givers. In ordinary conversation instances of troubles talk are recognizable as instances of heightened mutuality. In medical settings, patients demonstrate troubles resistance to show that they are not rushing to embrace the sick role, but rather that they recognize the appropriate roles and relationships of doctor and patient as professional and layperson. There is a convergence, then, between the resources available to patients and the techniques for establishing doctorability, and interactions between doctors and patients that establish doctorability also serve to produce and maintain normative roles and relationships (Gill, 1998).

It is important not only for patients but also for doctors that patients are able to present themselves and their complaints as doctorable. According to the definition of doctorability, patient accounts, including information transfer, will vary depending upon what doctors understand as constituting legitimate and treatable reasons for seeking their help, which will in turn be partly determined by doctors' ideas about their roles and relationships with their patients. Medical settings, furthermore, can be understood as organizational and spatial manifestations of varying types of medical understandings and practices as well as resources for the reproduction of these understandings and practices. Halkowski's (forthcoming) observation that doctors initially face the task of determining whether patient accounts are credible is related to the criteria for doctorability in conventional medical settings. In such settings, as Halkowski demonstrates, patient narratives of symptom discovery often function to "justify the patient's presence in the doctor's office with this problem" (Halkowski, forthcoming) in other words, they are one way of addressing the question of the 'reason for the visit' (Heritage, forthcoming) and corresponding issue of doctorability. Halkowski further notes that these narratives, as one of the ways that patients establish the reasonableness of their visits, can be thought of as one part of the broader patient "task" of "doing: 'being a good (i.e. competent) patient'" (Sacks, 1984) (Halkowski, forthcoming); that is, showing that one is attentive and responsive to situationally relevant doctorability concerns.

Information transfer in conventional medical settings takes place during interactions through which doctors seek to evaluate patient credibility and patients seek to demonstrate that they are indeed credible – i.e. "good" –

informants about their own bodies. Through the analytic lens that Halkowski provides, we can see how information transfer is influenced by the interactional goals of both doctor and patient (Halkowski, forthcoming), and I have argued further that these goals are shaped by variable criteria of doctorability. Information transfer will be a product of interactional collaboration between coparticipants and doctorability concerns will be present in encounters between doctors and patients across medical settings. The criteria for establishing a doctorable condition, however, differ across medical settings, and these differences are manifest in interactions between doctors and patients in the settings.

DOCTORABILITY AND THE DESCRIPTION OF BODILY EXPERIENCE

Halkowski's work illuminates part of the taken-for-granted normative order of conventional American medicine. Studying interaction in unconventional medical settings provides evidence for examining how orienting to different expectations and understandings (both social and technical) influences the style and content of doctor-patient communication. It also offers a perspective from which to interrogate the assumptions present in conventional encounters. The analysis of an unconventional medical interaction below, for instance, problematizes the assumptions that patients ought to be able to justify their visit to the doctor; that some types of accounts are more reliable than others; and, that conventional doctors are qualified to evaluate not just patient symptoms, but the veracity of patient reports.

Such questioning, moreover, points to important ways that understandings, technologies, institutional organization, and lay and professional identities, all constituents of what I refer to as medical settings, are present in concrete doctor-patient interactions and influence the production of descriptions of bodily experience. For example, the theoretical system underlying each type of medical practice, as understood by its practitioners, is an important determinant of what will be perceived by those practitioners as reasonable and relevant patient accounts. Because conventional American medicine is based on a biomechanical model of the body, practitioners are concerned with pinpointing the characteristics of particular symptoms with great accuracy.[6] According to Halkowski,

> Routinely, doctors have to make assessments of patients' "pain reports" in order to accurately prescribe medications. Thus the doctor constantly needs to have a sense of the patient's veracity as a reporter of bodily states, pain levels, etc. The patient's narrative of

> discovery is one piece of data which doctors employ (unreflectively, for the most part), to assess the veracity of patients

The technology of conventional medical pharmacology can thus be seen to be one source of the doctor relevance of judging the veracity of patient reports.

My research at the Healing Center supports the observation that when taking the role of patient individuals become concerned with "the task of doing: 'being a good patient.'" In other words, patients working to present themselves as reasonably in the doctor's presence with a doctorable problem can be observed in both conventional and unconventional medical settings. In addition, my research suggests that "the task of doing: 'being a good patient'" is accomplished in different ways in the different settings. I argue that this is related to differences in the doctorability criteria employed in the two kinds of medical settings examined in this paper. In my ethnography, for instance, I make the point that,

> Mainstream and Chinese medical practices draw different boundaries around the range of concerns that they consider to be within their sphere of authority and ability, and they use different criteria for matching client-patients with practitioners and treatments. For instance, based upon Chinese medical philosophy, the Chinese medicine practitioners recognize no distinction between the pathologies and treatments of physical and mental illness. This, in effect, expands the range of concerns that are considered reasonable in the setting of the Healing Center compared with the practice of a Western MD (Katz, 1996: 19).

I note, as well, that the practice of Chinese medicine in the U.S., especially among non-Chinese, tends to be restricted to non-emergency medicine and to specialize in the treatment of complaints that conventional medicine does not have adequate means to treat (e.g. chronic pain, stress, anxiety, fatigue). These circumstances, as well as the basic principles and pharmacopoeia of Chinese medicine which focus on treating the body's fundamental condition rather than on the treatment of specific symptoms, mean that there are few instances in which assessing patient credibility is a matter of importance in doctor-patient interactions at the Healing Center, and that there are few conditions that patients report that are not somehow doctorable.

Patients, accordingly, can be expected to choose to provide different information to doctors in different medical settings, based at least in part on their interactions with practitioners in these settings through which patients learn to produce descriptions of their bodily experiences that are doctorable. Interestingly, the norm of troubles resistance observed repeatedly in conventional medical settings seems to be relaxed in the unconventional setting. These observations suggest that the same patient may choose to report very different information in different kinds of medical interactions. Information transfer decisions, embedded in symptom reports, must then be understood to be learned and mutable orientations to specific types of doctor-patient interactions. Comparing doctor-patient interactions in conventional and unconventional medical settings thus helps to make observable the social processes involved in constructing the content and form of patient symptom reports. In the discussion that follows I will illustrate some of the differences in patient accounts in conventional and unconventional settings and discuss how they are part of the way that patients orient to "the task of doing: 'being a good patient'" in these settings.

THE INTERACTIONAL PRODUCTION OF SYMPTOM REPORTS

(a) Patient Contributions

Halkowski (forthcoming) identifies a set of communicative devices that patients in conventional settings employ to produce narratives of symptom discovery. He then analyzes these narratives to show how they function both to arrive at an understanding of what the problem is and at the same time to address reliability and doctorability concerns. For instance, he observes "doing: being uncertain" to be one way in which patients' try to establish their credibility as witnesses of their bodies. Being uncertain, Halkowski notes, comprises at least three aspects: unintentionality, approximateness, and passiveness. That is, by being uncertain patients indicate that have discovered the problem "accidentally" and thus display themselves as being reasonably, but not overly, careful monitors of their bodies. When the patient treats his observational activities as accountable – i.e. requiring explanation – his credibility is affirmed because he shows that he is aware that in this context it is considered abnormal to be a very careful observer of one's body. In the transcript excerpt included below,[7] note how the patient uses temporal formulations, qualifiers, and a passive stance to present himself as having discovered the symptom accidentally and as "having pursued medical care for this problem reasonably quickly" (Halkowski, forthcoming).

(1) Medical Resident (M) talking with Patient (P)

P: I do have one thing I want you folks to look at today that, that's come
up real . . . since I saw doctor Bergman last . . . matter of fact I only
noticed it about ahh . . . I'm gonna say two weeks, or . . .
M: Okay,
P: three weeks ago.

P: I have ah- ah little swelling here.
M: Okay,
P: and ah, I don't know whether it's ah, ah hernia, or something inside there causing it, but ah, it is, ah little lop sided maybe I'm just growing that way.
M: Hm hmmm,
P: But I think maybe it's something ought ah be looked at.
M: You just noticed it two weeks ago,
P: Yeah.
M: Okay.
P: It coulda been there for ah year. I don't look at myself very much, you know.
M: I see.
P: But I was shaving or something and I do some side to side exercises an I guess I was doing it an kindah maybe in front of ah mirror or something an I just noticed that this . . .
M: hm hmmm,
P: side is extended
M: Okay.
P: rather than this side.

Looking more closely at Halkowski's (forthcoming) analysis, first, the temporal formulations, taken together, emphasize on the one hand uncertainty ("I'm gonna say two weeks, or . . . three weeks ago," "It coulda been there for ah year") and on the other hand reasonableness, by suggesting that the patient did not engage in any extended period of self-monitoring before seeking the advice of a physician. The doubt with which the patient reports the timing of symptom discovery can also be taken to indicate the disinterestedness or absence of intentionality characteristic of taking a passive stance toward symptom discovery. According to Halkowski (forthcoming), such finely balanced reporting is a way that patients in conventional settings attend to the task of presenting themselves as responsible witnesses of their bodies.

Second, the qualifiers used by the patient further serve to represent the patient as "uncertain." Halkowski (forthcoming) divides these linguistic devices into two types: approximators (e.g. "or something") and mitigators (e.g. "I guess," "maybe," "kindah"). Note how in lines 15–16 the patient uses approximators both when presenting his concerns (that it might be "ah hernia or something") and when describing how he first noticed the problem ("I was shaving or something"). He uses mitigators in lines 18 and 23, often along with approximators, whenever reporting on aspects of the problem and its discovery that might be diagnostically relevant. That this is not a personal idiosyncracy is suggested by the very assertive statement that "I don't look at myself very much, you know," which indicates that the patient is concerned with presenting himself as a certain kind of monitor of his body. In other words, the patient orients to being a disinterested monitor of one's body as normal and expected. By being approximate or uncertain the patient leaves statements of precision and certainty to the physician, in effect, distributing responsibility for the identification of diagnostically meaningful information and of the condition itself to the doctor.

Finally, the patient takes a passive stance towards his symptom discovery (Halkowski, forthcoming). The patient uses two formulations in his opening statement that clearly establish that he was not actively looking for any problems with his body: the first when he refers to his concern as something "that's come up," and the second when he adds, "I only noticed it about . . ." The absence of a subject in the phrase "that's come up" presents the bodily change as having imposed itself upon the observer, and the verb 'to notice,' by definition, conveys the idea of finding something one was not consciously looking for. Moreover, the entire narrative structure of the account functions to present the patient as a passive monitor of his body who just happens to notice, in the course of doing other things, that he has a problem (Halkowski, forthcoming).

This piece of data also includes examples of troubles resistance. While the decision to visit a doctor itself suggests the judgment that one's problem merits medical attention, reporting the ways that one has resisted arriving at that decision is used to uphold the soundness of that judgment (Heritage, forthcoming). Some ways of showing troubles resistance are making normalizing or minimizing statements and employing an "objective, 'just the facts' approach to illness" (Heritage, forthcoming). The patient can be seen to minimize his condition by identifying it as "ah little swelling here" (line 6) and to attempt to normalize it by commenting that "maybe I'm just growing that way (line 9)." In addition, the patient mostly employs a 'just the facts' manner of reporting, omitting any subjective comment on his experience of, or concerns about, his condition.

If doing "being a good patient" means basically the same thing in every doctor-patient interaction, then we should see patients using similar temporal and linguistic formulations and orienting similarly to agency in the conventional and unconventional settings. Differences in the formulations and orientations, accordingly, would indicate differences in the normative understandings and doctorability criteria of the two types of settings. That patients in the unconventional setting can be seen to use distinctly different communicative devices and to construct distinctly different accounts from those in the conventional setting suggest the latter. Consider the following example from my field notes.

(2) Patient (P) talking with Intern[8] (I)

P: Hi, how are ya?
I: Fine. How can I help you?
P: I need more of this, but this gave me strange effects. First it made me constipated, and then it went to the other end of the spectrum. Really mushy and disgusting. Since you guys always ask about consistency, I thought I'd just tell you everything.
I: How do you feel now? (He starts to check her pulses.)
P: Okay. Almost normal, but it's still a little soft.
I: Let me run through this. First you got constipated and then you had diarrhea . . .
P: Now it's mushy and regular, but mushier than before I started taking this stuff.
I: Does your tongue look any different? (He starts to pull down some bottles from the shelves behind him as he listens)
P: I can't really tell. I brush three times a day but I don't really know if it looks any different. (The intern asks to see her tongue)
I: Wow. Three times a day. That's great.
P: Well, in my business I have to be near people and I want to have clean breath. I've heard it's good to brush your tongue. I had some lower back pain from the constipation, and then my head started to hurt.
I: Did you have any problems before taking these herbs?
P: Not with elimination.
P: One other thing, I have no appetite. I didn't mean to hold it back from you, but I just thought of it now.
I: As a result of these formulas?
P: I think so. I had an appetite before.
I: OOOOh! (He immediately rises and dashes back to the shelves and grabs a bottle of pills.) This will stimulate your appetite and should take care of the rest of it. Let's get rid of these other ones for now. Just take this one.

First, note the orientation to time. The symptom report in the conventional setting (1) is built around a narrative of discovery that emphasizes uncertainty and accidental discovery. In contrast, the symptom report above (2) is characterized by 'sequential/causal' formulations ("first . . . then," "before," "as a result") that emphasize just the opposite: these formulations demonstrate the patient's intentional observation of changes in her body's condition and her body's responses to various treatments. That this form of reporting is part of the task of doing 'being a good patient' is suggested by the account in lines 5–6 ("Since you guys always ask about consistency . . .") where the patient states explicitly that she is providing information that she has learned is doctor relevant in this setting. It is even more strongly indicated by lines 22–23, where the patient holds herself accountable for not telling the doctor everything up front ("I didn't mean to hold it back from you . . ."). In this interaction, the patient sees her task as providing as much detail about her condition as possible. In the way she does this we can also see a relaxation of the norm of troubles resistance. The doctor relevance of patients sharing 'everything' with their doctors encourages inclusion of the subjective commentary found in troubles telling (e.g. line 4: "Really mushy and disgusting"). Contrary to Jefferson's (1981) findings, in the unconventional setting communicative acts between doctor and patient that move towards establishing mutuality appear unproblematic and even welcome. This difference is striking and itself warrants further investigation. In the unconventional setting the patient aims to demonstrate her competence by showing that she is a diligent and conscientious monitor of her body. I will call such accounts by patients in the unconventional setting monitoring reports, which may include narratives of symptom development.

The relevance of reports of symptom development, but not of symptom discovery, in this setting may be partly attributed to Chinese medical principles, which focus on treating the current condition of the patient, understood as a moment in a continually changing system. According to this way of understanding health, human bodies always manifest symptoms of change, owing to the integrated body/mind system continually adjusting to changes in its internal and external environments. Chinese medicine's methods of diagnosis, as observed at the Healing Center, include the doctor continually asking for, and attending to, patient descriptions of a range of symptoms. This practice has at least two important interactional consequences. One is that it distributes responsibility for arriving at a diagnosis between the doctor and patient; another is that the patient is encouraged to come up with additional information.[9] In the above example, this process leads to the introduction by the patient of a new piece of information, lack of appetite (line 23), that is oriented to by the doctor as holding the key to the diagnostic puzzle. As noted above, this addendum to the patient's monitoring report indicates that, when orienting to the normative order of the Healing Center, it is not providing, but withholding, information that requires an account. Thus, "the task of doing:

'being a good patient'" in this setting includes transferring as much information, from the broad range of doctor relevant information, as possible, as well as responsibility sharing between doctor and patient.

A second contrast between the interactions in the two settings is the use of qualifiers. Whereas in the conventional setting the patient makes use of approximators and mitigators to project uncertainty and disinterestedness, in the unconventional setting not only are approximators and mitigators absent, but the patient uses devices that I will call specifiers ("before," "first") and intensifiers ("really," "much"), which create the opposite effect of certainty and interest. Here we see that the patient presents herself as being observant and knowledgeable, and she uses linguistic devices that project and affirm this. For instance, there is no equivocation by the patient about the consistency of her stool. Based on normative attitudes towards elimination, we might expect a person to be uncertain about the texture of her stool. But, this patient not only tells the doctor that her stool was "really mushy and disgusting" but is also able to evaluate that it is "Almost normal, but it's still a little soft" indicating that she observes her stool regularly enough to know what is and isn't "normal." Moreover, note the confidence of the patient's responses throughout the interaction. Whenever the doctor asks a question, the patient provides a straightforward answer ("Yes," "No," "I can't really tell"). What is important is that in the unconventional setting it is oriented to as "normal" to be a regular and careful monitor of one's body and thus another strategy for accomplishing "the task of doing: 'being a good patient'" is to be certain about one's observations of one's body.

Third, the patient in this example demonstrates an active stance towards self-monitoring. Both the intentionality of the sequential/causal formulations and the certainty with which the details of the report are delivered contribute to demonstrating an overall active orientation to self-monitoring on the part of the patient. Here, there appear to be no limits to how careful and deliberate an observer of her body the patient can be and still maintain credibility.

While patients in the unconventional setting often appear eager to demonstrate that they have been good self-monitors by providing exhaustive monitoring reports, patients in the conventional setting face the more demanding task of maintaining a fine balance between showing themselves to be reliable monitors, while avoiding the impression that they observe themselves with too much frequency or interest. Indeed, in the example from the conventional setting, the latter part of the interaction is shaped by the doctor's sanctioning of the patient (line 15) for being too lax an observer of his body. By ignoring the patient's candidate diagnosis (lines 9–11) to focus on time of noticing, the doctor makes clear that the patient's credibility is still an issue. This interpretation is supported by the videotape of this interaction in which it is quite obvious that this is no minor swelling that could be easily missed.

The evidence from the conventional interaction suggests that the doctor's concern with the trustworthiness of this patient's symptom report may be warranted. However reasonable, emphasizing credibility concerns over attending to the problem itself and the patient's candidate diagnoses still functions to limit information transfer. One consequence of ignoring the patient's candidate diagnoses is that the diagnostic labor remains solely the responsibility of the physician. Such disattending to patient contributions may be understood as a way in which physicians establish boundaries and claim control of diagnoses, and it may also serve patient goals of being relieved of responsibility for the illness and its cure.[10] This is also a good example of how patients learn what is doctor-relevant from the questions that doctors ask. In this case, the patient learned that, at least until he proved himself a reliable informant, temporal noticings were relevant while candidate diagnoses were not.

(b) Doctor/Practitioner Contributions

Although the aim of this paper is to focus attention on the work that patients must attend to in doctor-patient interactions, and how the nature of this work varies by type of setting or interaction, it will be further illuminating to turn our attention briefly to the role of the doctors in these two settings. This will also further elucidate how the construction of the patient role as passive or active is a collaborative product of the doctor-patient interaction.

How does the doctor in the conventional interaction (1) participate in the construction of the patient's symptom report? As discussed above, the doctor's orientation to evaluating the patient's reliability as an informant impacts on the content of the patient's report. One way this is done, as we saw, is to focus on the patient's competence as an observer of his body – specifically, in this case, regarding when the patient noticed his symptom. Other than this reference to time of noticing (line 15), all of the doctor's contributions are what are known as continuers, words and sounds ("okay,", "hm,hmmm," "I see") that serve to show that the doctor is listening and that the patient may continue. They reveal nothing about the state of mind of the listener, but simply permit further elaboration. Use of continuers by doctors in doctor-patient interactions is one way that asymmetry between doctor and patient is interactionally produced (ten Have, 1991).

In contrast, in the unconventional setting the practitioner immediately focuses on the patient's problem. There is no evaluating of the patient's credibility. The practitioner quickly begins (line 7) asking questions directed specifically at the patient's condition ("how do you feel now?"), and attends immediately (line 9) to the content of the patient's response ("Let me run through this . . .") We can see revealed in the doctor's questions an implicit assumption that patients not only monitor their conditions but that they are (or at least can be) highly competent monitors. For instance, when the doctor asks, "Does your tongue look any different?" he is orienting to the patient as being potentially knowledgeable about the method of tongue diagnosis used in Chinese medicine.

In terms of the doctor's responses to the patient's reporting, if I had collected recorded data my transcripts might contain more continuers than it does at present (the only way to find out would be to collect such data). Nonetheless, I record the doctor using two types of responses that would be highly unusual in a doctor-patient interaction in a conventional setting: "wow" and "oh." These two responses, in particular the use of "oh" have been called change-of-state tokens (Heritage, 1984) and indicate that the listener knows something new or different because of what the speaker has just said. It has been shown that part of "doing: 'being professional'" involves refraining from using such responses with clients (ten Have, 1991). The use of such responses in the unconventional setting suggests a qualitative difference in the relationship between doctor and patient, in particular regarding the distribution of responsibility and authority. This interpretation is supported by an explicit, unsolicited statement made by one of the practitioners at the Healing Center:

> For me the doctor-patient relationship is a contract. The person has to agree to work together with me because Chinese medicine is a whole lifestyle thing. You can't just come in here and say "cure me." If someone has an ulcer, for example, we can give them herbs and acupuncture in here, but if they go back out there and do what they were doing before it's not going to make any difference. It's got to be an equal relationship (Katz, 1996: 22).

This quote also offers a glimpse of a way in which the normative order oriented to in the unconventional encounter differs from that in the conventional encounter. Notably the practitioner's view of the contractual and equal relationship between doctor and patient. In the conventional setting, and as Parsons prescribed, the patient must give up his/her responsibility to the doctor in order to receive treatment. According to this practitioner, the situation in the unconventional setting is just the reverse: the patient must agree to take responsibility in order to receive help. What the unconventional practitioner calls 'an equal relationship' is in many ways similar to what conventional practitioners call 'compliance,' yet the difference represented by the choice of terminology is not trivial. It is another illustration of how doctor and patient roles and relationships are constructed differently in the two kinds of settings.

CONCLUSION

This chapter has begun an exploration of how "the task of doing: 'being a good patient'" comprises different ways that patients orient to and talk about changes in their bodies in different types of settings, and how these differences in the structure of attention and expression have implications for the co-construction of roles, relationships and information transfer. The main differences observed in the interactions in the conventional and unconventional settings are summarized in Table 1. In the conventional setting, data segment (1), the patient orients to "the task of doing: 'being a good patient'" by presenting himself as a casual and disinterested observer of his body; by producing a narrative of discovery that is also an attempt to assert his reasonableness and credibility; and by treating his self-observation and visit to the doctor as accountable. This is done through the interactional deployment of a range of communicative devices, by both patient and doctor, that concurrently serve to distribute diagnostic responsibility to the doctor and to limit information transfer by patient. Differently, in the unconventional setting, data segment (2), the patient orients to the task of doing: 'being a good patient' by presenting herself as a careful and intentional observer of her body; by producing a confident, detailed monitoring report of her symptom development; and by treating omission of information as accountable. The communicative devices deployed in this interactional process promote shared responsibility for diagnosis and encourage information transfer between patient and practitioner.

The issue of doctorability has been shown to influence both the content and style of patient accounts, as well as doctor and patient roles and relationships. The linking of doctorability with evaluation of patient credibility in the conventional setting leads to an asymmetrical doctor-patient relationship and limits information transfer. In contrast, the orientation to equality and responsibility sharing in the unconventional setting promotes mutuality and information transfer in the doctor patient relationship. By closely examining the communicative devices used in two doctor-patient interactions, this paper has demonstrated how, even when information transfer is understood as influenced by doctorability concerns, patients participate actively by doing the work of deciding what information is doctor relevant and then presenting it in the manner they understand as most appropriate to the "task of doing: 'being a good patient'" in the particular setting. Finally, constructing doctor relevant accounts has been shown to have both technical and social dimensions – that acts of giving and taking information are both moves toward diagnosis of an illness as well as moves toward establishing and maintaining roles and relationships. Consequently, an adequate understanding of information transfer decisions must take into account the social dimensions of the doctor-patient encounter.

Table 1. Interactional Constituents of Constructing the 'Good Patient' During Symptom Reporting in a Conventional and an Unconventional Medical Setting

Constituent of Interaction	Conventional Setting	Unconventional Setting
orientation to self-observation	patient orients to being a casual, disinterested and accidental monitor of one's body as normal	patient orients to being a diligent and deliberate monitor of one's body as normal
orientation to doctorability	patient must demonstrate a credible problem and reason for visit	all patient concerns and reports treated as credible
treated as accountable	noticing bodily functions and changes; visiting the doctor	omitting information about bodily functions and changes
communicative devices, patients	time-when formulations; approximators and mitigators; taking a disinterested, passive stance towards symptom discovery; troubles resistance; narratives of discovery	sequential/causal formulations, specifiers and intensifiers; taking an interested, active stance towards self-monitoring; absence of troubles resistance; monitoring reports
communicative devices, doctors	continuers – serve to preserve distance between doctor and patient; maintains doctor's professional reserve	change of state tokens – serve to narrow distance between doctor and patient; establishes egalitarian relationship
division of diagnostic labor	interaction distributes to doctor responsibility for diagnosis	interaction distributes diagnostic labor between doctor and patient
information transfer	interaction encourages patient withholding of information and keeping talk to 'just the facts'; part of "being a good patient" is carefully selecting information to share with doctor	interaction encourages patient talk about the body and self-care practices; reporting a wide range of information is encouraged and is part of how one demonstrates being a "good" patient

Admittedly, the issue of patients withholding information specifically about their use of unconventional therapies from their conventional doctors can not be directly addressed with the data presented here. The data do suggest, however, that the idea of patients refraining from talking about unconventional therapies with their conventional doctors because they suspect that such information could jeopardize their credibility, and thus affect the quality of their conventional medical care, is worth considering. This would not preclude other explanations. Patients might, for instance, withhold information because they themselves are ambivalent about the unconventional therapies or because they don't see their use of unconventional therapies as relevant in conventional settings. Also, that patients in the unconventional setting are encouraged to talk more and to establish more equal relationships with their doctors than in the conventional setting does not mean that doctors in these settings are necessarily more altruistic than doctors in conventional settings. Thus it is also important to understand the ways that the orientations and interactional patterns observed in the unconventional setting could be manipulated or misused, for instance to minimize practitioner accountability and maximize treatments and product sales.

Without minimizing these concerns, the widely recognized problems in doctor-patient relationships in conventional settings, where asymmetry is the norm, make further research on types of patient-provider relationships that emphasize mutuality important. Additionally, if there is interest in applying the findings of such research to the transformation of doctor-patient interaction in conventional settings, the following questions inevitably emerge: Under what conditions does establishing greater mutuality between doctors and patients also facilitate the construction of mutually acceptable roles and relationships; and, what would be the necessary prerequisites for and consequences of changing conventional styles of doctor-patient interaction? Since the work that patients do, in collaboration with doctors, to accomplish "the task of doing: 'being good patients'" has been shown to have important implications for information transfer and diagnosis as well as for the construction of roles and relationships, the value of answers to these questions will depend on how well they address the mutual embeddedness of technical and social processes.

NOTES

1. Notable exceptions are Gill (1998), Halkowski (forthcoming), and Heritage (forthcoming).

2. Concern regarding a deficiency in doctor-patient relations is not new (Balint, 1957; Szaz & Hollender, 1956) and has paralleled the rise and consolidation of scientific medicine, and the corresponding segregation and marginalization of the

psychosocial dimensions of illness and healing (e.g. see Shorter, 1985). What is new is how the changing institutional environment is changing the experience of doctors within the medical system.

3. I discovered that my use of the concept of "setting" became problematic when drawing comparisons with work coming from a conversation analytic perspective. From that perspective, interaction is a more robust and accessible phenomenon than setting, as well as a fundamental constituent of setting. At the risk of oversimplifying, conversation analysts are primarily interested in interactions (as the fundamental building blocks of social life), whereas ethnographers are primarily interested in sites (places where relatively regular and structured social activities takes place) and how all of the dimensions of the site that are accessible to their methods of collecting data (e.g. social organization and procedures, informal social groups, ideologies, physical environment, personal and organizational biographies, etc.) interact in the local production of meaning. Accordingly, the concept of setting has been central to my own work. My original study was of a specific place and the interactions that occurred there, and my task was to weave together the richly textured information I gathered through observation, participation, field-based and stand alone interviews, and reading of promotional and educational materials to construct my analysis. So, while I agree that interaction is an important constituent of setting, and that in many places in the paper it would be equally acceptable to replace the term 'setting' with the term 'interaction,' I have decided, for the above reasons, to keep the term 'setting.'

4. I call this type of medicine 'Chinese medicine' rather than the now widely used 'traditional Chinese medicine' or TCM.

5. Hence, perhaps, current interest in a patients' "Bill of Rights."

6. Thus, when Western-trained doctors do share their thought processes with their patients, it is most likely to be in the form of a matching of symptoms with diagnoses and the treatment options associated with these diagnoses. They tend to provide description rather than explanation. In Chinese medicine, in contrast, symptoms tend to be explained with reference to the theoretical system, which then points to possible diagnoses and treatments.

7. To facilitate comparison, I have simplified the transcript conventions. For the original, fully detailed, transcript, see "Realizing the Illness: Patients' Narratives of Symptom Discovery" forthcoming in *Practicing Medicine: Talk and Action in Primary Care*, edited by John Heritage and Douglas Maynard.

8. Interns at the Healing Center were students in the final stages of their degree programs at a local college of oriental medicine.

9. It is interesting to compare this with Beckman and Frankel's (1984) finding that the communication style of physicians in conventional settings commonly "inhibits further patient identification of additional concerns" and leads to "the loss of patient information" (Beckman & Frankel, 1984: 694).

10. The patient here may be contributing by presenting information in a way that is ignorable, a view which is suggested by the work of Virginia Gill (1998) that examines patient explanations as part of the way "patients display and doctors ratify a precisely circumscribed scope of lay knowledge regarding illness," a perspective that shifts attention away from doctor control to the "collaborative activities whereby patients and doctors construct and preserve their identities as laypeople and professionals."

ACKNOWLEDGMENTS

The strengths of this chapter owe much to the careful reading and thoughtful comments provided by my professors at UCLA, John Heritage and Robert Emerson; the attendees at the Pacific Sociological Association panel where I presented a much earlier version of this paper; the editor of this series, Jennie Jacobs Kronenfeld; and especially Timothy Halkowski, who is to be commended for his generosity of feedback in light of a graduate student's struggle to find her own voice (without having to distort someone else's in the process). In addition, Tiffani Chin, Katherine Hill, Maggie Kusenbach, and Jennifer Leich provided very useful suggestions and invaluable support. I also want to thank Steve Clayman for his help and encouragement.

REFERENCES

Balint, M. (1957). *The Doctor, his Patients and the Illness*. London: Pitman Medical Publishing Co.

Becker, H. S. et. al. (1961). *Boys in White: Student Culture in Medical School*. Chicago: University of Chicago Press.

Beckman, H., & Frankel, R. (1984). The effect of physician behavior on the collection of data. *Annals of Internal Medicine*, *101*, 692–696.

Byrne, Patrick and Barrie Long (1976). *Doctors Talking to Patients*. London: Her Majesty's Stationery Office.

Eisenberg, D. M. et al. (1993). Unconventional medicine in the United States: Prevalence, costs, and patterns of use. *New England Journal of Medicine*, *328*(4), 246–252.

Frankel, R. (1990). Talking in interviews: a dispreference for patient initiated questions in physician-patient encounters. In: G. Psathas (Ed.), *Interaction Competence*. Lanham, MD: University Press of America.

Freidson, E. (1970). *Profession of Medicine: A Study of the Sociology of Applied Knowledge*. Chicago, University of Chicago Press.

Gill, V. (1998). Doing attributions in medical interaction: Patients' explanations for illness and doctors' responses. *Social Psychology Quarterly*, *61*(4), 342–360.

Halkowski, T. (forthcoming). Realizing the Illness: Patients' Narratives of Symptom Discovery. In: J. Heritage & D. Maynard (Eds), *Practicing Medicine: Talk and Action in Primary Care*. Cambridge: Cambridge University Press.

Heritage, J. (1984). A change-of-state token and aspects of its sequential placement. In: J. M. Atkinson & J. Heritage (Eds), *Structures of Social Action: Studies in Conversation Analysis*. Cambridge: Cambridge University Press.

Heritage, J. (forthcoming). Accounting for the visit: Giving reasons for seeking medical care. In: J. Heritage & D. Maynard (Eds), *Practicing Medicine: Talk and Action in Primary Care*. Cambridge: Cambridge University Press.

Jefferson, G. (1981). The Rejection of Advice: Managing the Problematic Convergence of a "Troubles-Telling" and a "Service Encounter." *Journal of Pragmatics*, *5*, 399–422.

Jefferson, G. (1988). On the Sequential Organization of Troubles-Talk in Ordinary Conversation. *Social Problems*, *35*(4), 418–441.

Katz, M. (1996). The Dialectics of "Alternative"/"Mainstream" in the Practice of Chinese Medicine in the United States. Unpublished Manuscript.

Light, D. (1980). *Becoming Psychiatrists: The Professional Transformation of Self.* New York: Norton.

Mishler, E. (1984). *The Discourse of Medicine: Dialectics of Medical Interviews.* Norwood, NJ: Ablex.

Parsons, T. (1951). *The Social System.* New York: Free Press.

Parsons, T. (1975). The sick role and the role of the physician reconsidered. *Milbank Memorial Fund Quarterly, 53*, 257–278.

Roter, D. L., & Hall, J. A. (1992). *Doctors Talking with Patients/Patients Talking with Doctors.* Westport, CT: Auburn House.

Sacks, H. (1984). On doing: "Being ordinary." In: J. M. Atkinson & J. Heritage (Eds), *Structures of Social Action.* Cambridge: Cambridge University Press.

Shorter, E. (1985). *Bedside Manners: The Troubled History of Doctors and Patients.* New York: Simon and Schuster.

Szaz, T., & Hollender, M. (1956). The Basic Models of the Doctor-Patient Relationship. In: H. D. Schwartz (Ed.), *Dominant Issues in Medical Sociology* (1987, 2nd ed.). New York: Random House.

Ten Have, P. (1991). Talk and institution: a reconsideration of the "asymmetry" of doctor-patient interaction. In: D. Boden & D. Zimmerman (Eds), *Talk and Social Structure* (pp. 138–63). Cambridge: Polity Press.

PREVENTING HOSPITALIZATION: HOME HOSPICE NURSES, CAREGIVERS, AND SHIFTING NOTIONS OF THE GOOD DEATH*

Jennifer Leich

ABSTRACT

This chapter explores the relationship between hospice home care and the recent trend toward dehospitalization. Hospice home care nurses are increasingly responsible for attending to their employing organization's bottom line, in addition to patient care. Their concern for helping patients achieve the good death thus intersects, and sometimes competes, with their employing organization's interest in controlling patients' access to costly hospital and other acute care. I describe several of the ways in which nurses in this study control home hospice caregivers' abilities to support terminally ill patients' death at home: (1) through their assessments of patient and caregiver appropriateness and competence; (2) by appealing to the caregiver's emotional connection to the patient in order to encourage commitment to the goal of death at home and; (3) by normalizing the idea of death at home. I argue that, through their work of training hospice caregivers to adapt and maintain the value of staying at home, hospice home care nurses provide a subtle, yet significant, new

* An earlier version of this chapter was presented at the Society for the Study of Social Problems Annual Meeting, August 1998, in San Francisco, CA.

Health, Illness, and Use of Care: The Impact of Social Factors, Volume 18, pages 207–228.

ISBN: 0-7623-0740-4

mechanism for dehospitalizing terminally ill patients. Moreover, the work of hospice home care nurses is shifting our understanding of what constitutes a "good death".

INTRODUCTION

This chapter explores the relationship between hospice home care, which has traditionally focused on patient-centered care delivery processes, and the recent trend in healthcare toward dehospitalization. An important role of the hospice nurse is to facilitate the terminally ill patient's "good death." Hospice home care nurses, like all home care nurses, are increasingly responsible for attending to their employing organization's bottom line, in addition to patient care (Dee-Kelly, Heller & Sibley, 1994). What this means in the case of hospice home care is that nurses' concern for helping patients achieve the good death intersects, and sometimes competes, with the agency's interest in controlling patients' access to costly hospital and other acute care. I argue in this chapter that, through their work of training hospice caregivers to adapt and maintain the value of staying at home, hospice home care nurses provide a subtle, yet significant, new mechanism for dehospitalizing terminally ill patients.

By trying harder to control patients' (and their informal caregivers') use of health care services, current hospice home care delivery is changing the emphasis that hospice has tended to place on patient autonomy. Traditionally in hospice, patient autonomy – a key part of what is thought to constitute the good death – has been synonymous with patients' ability to assert their wishes to remain at home (and out of the hospital) for end of life care. In today's managed care environment, and with the growing acceptance of hospice among insurers and healthcare providers, autonomy struggles often center on whether or not patients can gain access to inpatient hospital care (Gordon, 1997). Several factors may influence patients' and families' desires to use hospital care in conjunction with hospice: uneven quality and fragmentation of hospice care; a frequent lack of the comprehensive support that the hospice philosophy calls for; lack of acceptance of the hospice philosophy on the part of patient or caregivers; variability in families' caregiving abilities and; situational factors. Once patients sign up for hospice, however, they are channeled away from using hospital care.

The ways in which current hospice home care practice may undermine patient autonomy show up when conflicts arise over exactly how and where patients wish to die. Patients' and caregivers' attempts to control the care patients get, including their decisions about whether or not to use hospital care, can place them at odds with home hospice organizations, whose goals increasingly include limiting patients' use of inpatient care.

Hospice has grown in scope and importance in part because of its perceived potential to limit costly care at the end of life (Greer, Mor & Kastenbaum, 1988a; Mor & Kidder, 1985). While the research evidence on hospice's cost-effectiveness has been mixed, studies that do document reduced costs point to decreased hospitalization as a major factor in the savings (Hughes, Ulasevich, Weaver, Henderson, Manheim, Kubal & Bonarigo, 1997; Hughes, Cummings, Weaver, Manheim, Braun & Conrad, 1992; Kidder, 1988; Mor & Kidder, 1985; Brooks & Smyth-Staruch, 1984). As hospice organizations use the promise of reduced costs to legitimize their place in the mainstream health care system (Goodlin, 1997; Mahoney, 1997), others worry that its original mission may be getting lost in the process (Beresford, 1997; McNamara et al., 1994; James & Field, 1992; Abel, 1986).

There is some concern that research evidence for hospice's cost-effectiveness could be due to selection bias in current studies rather than an actual "hospice effect" (Kidder, 1988; Brooks & Smyth-Staruch, 1984). In particular, the assumption is that people who self-select for hospice may be less predisposed to use intensive treatment modes than those patients who do not choose hospice. As yet there is no strong research evidence addressing this question.

My field observations of hospice home care suggest that both choosing hospice care and decisions about subsequent service utilization are much more complex, socially mediated processes than the prevailing self-selection model implies. Upon receiving a terminal diagnosis, some people have strong ideas about the kind of care they want and actively seek out hospice care. In other cases, patients and their families receive information, from a variety of sources, about the various options for treatment, including hospice. This information comes through a range of involved actors with potentially competing interests – the patient's caregiver and/or family members, friends, clergy, etc., the referring physician and specialists, the hospice agency, and the patient's insurance carrier. Moreover, the decision to enter hospice is not necessarily reached prior to the initiation of care, nor is it static. Typically, this decision is arrived at tentatively and can often change during the course of care. Decisions regarding both election of hospice benefits and utilization of services may be initiated by people other than the patient, and are subject to local, immediate pressures and constraints specific to the health care provider and/or insuring organizations within which the patient is embedded. In short, decisions about whether and how to use hospice are part of a fluid, negotiated, and oftentimes contested social process.

An analysis that acknowledges a wider range of actors and a broader scope of information involved in hospice selection and utilization allows us not only to understand more fully which factors predispose people to hospice in the first place, but also to understand what factors affect both utilization and outcomes. This latter issue relates directly to the question I am concerned with in this paper, namely – how are home hospice patients' and caregivers' decisions regarding utilization mediated by health care institutions and professionals? Specifically, I aim to show how and why home hospice nurses may influence caregivers' decisions to use – or not to use – hospital care in the final stages of patients' terminal illness.

METHODS AND STUDY PARTICIPANTS

Data and Method

I gathered the data for this case study using ethnographic field methods (see, for example, Emerson, Fretz & Shaw, 1995). Between February and November 1997, I observed the work of hospice home care nurses at the Home Hospice Association (HHA),[1] a large home hospice agency in Southern California. I observed nurses' daily round of work activities including case management, home nursing visits and case conferences. Five of the agency's ten nurses agreed to have me accompany them on patient visits and I traveled with four, making a total of twenty-seven patient visits. I also conducted in-depth, loosely structured interviews with 5 nurses. Three of the nurses were interviewed once and the other two were interviewed twice. Interviews lasted from one and a half to three hours.

The fieldnote and interview data that form the basis of my analysis focus on the interactional and decision-making processes that make up nurses' work with home hospice patients and their caregivers. I concentrate on work processes in order to investigate how the institutional context within which nurses work shapes the care they provide. In other words, I view work processes as a way to understand how organizational structure affects interaction, viewing work processes as phenomena that can be "dealt with as structures . . . which will exhibit a multitude of consequences" (Schatzman & Strauss, 1973: 6).

Study Participants and Setting

Of the nurses who participated directly in the study, three had over ten years experience as home hospice nurses and the other two had less than two years.

Two of the nurses worked full time (5 full days a week), carrying patient loads of 8–10, while the other two nurses worked part-time (3 full days a week) and carried patient loads of about 4–6 patients each.

HHA's hospice interdisciplinary team includes 10 registered nurses, two physicians, two medical social workers, one interdenominational chaplain, one hospice clinical supervisor, one assistant supervisor, one administrative secretary, one volunteer/bereavement coordinator, and a pool of hospice volunteers. The interventions of all (paid) disciplines must justifiably contribute to the support of the patient's medical management in the home. Most of HHA's ten hospice nurses work full time. All HHA nurses are paid on a per visit basis. Some have specialized training in hospice while most have trained and worked in other specialties before coming to HHA, where they learned their hospice skills on the job. Most, but not all, have had home health nursing experience before becoming hospice home care nurses. All HHA nurses work exclusively as hospice home care nurses.

HHA nurses act as "nurse/case managers" for their patients, which means that they are charged with managing all aspects of each of their patients' cases. They are accountable to the nursing supervisor and expected to have a wide knowledge of the agency's policies and procedures. All nursing visits take place in patients' homes on an appointment basis. While each visit may vary in terms of the focus of care and available time, nursing visits typically last about 45 minutes and include a thorough physical exam to assess patients' pain level and general physical condition. Nurses track the progression of patients' disease process and related physical condition, treating pain and symptoms as needed. In concert with others on the interdisciplinary team, they create the patient's hospice plan of treatment. Nurses assess caregivers' and patients' compliance with the hospice plan of treatment on an ongoing basis, making adjustments as needed and educating patients, caregivers, and family members regarding the death and dying process, pain control, symptom management and general patient care regimes. In addition, nurses provide supportive counseling and a compassionate presence for patients, their caregivers, and other involved family members and friends.

The ethnographic data I gathered for the study also include observations of the activities of hospice primary caregivers, most of whom were family members of hospice patients. In some cases, the bulk of direct patient care was provided by either trained or untrained caregivers employed directly by the patient or the patient's family. Hospice caregivers provide custodial and direct, "hands-on" patient care such as feeding, bathing, toileting, diapering, and transferring patients. Caregivers must monitor and deliver complex medications regimes, oftentimes doing tasks that even a licensed vocational nurse is not qualified to do in the home, such as administering narcotics. In addition, they are sometimes responsible for activities such as cleaning open wounds, catheters, and para-enteral ("tube") feeding equipment, all of which require acute attention to sanitary procedure.

THE GROWTH AND SIGNIFICANCE OF HOSPICE HOME CARE

While hospice home care constitutes a relatively small proportion of total U.S. health care spending, its importance is growing. Between 1988 and 1995, Medicare's expenditures for hospice grew from $118.4 million to $1.8 billion (NHO, 1999).[2] This growth is largely due to hospice's perceived potential to cut health care costs. The hospice philosophy began to take hold at a time of heightening concern about the escalating costs of health care. Early support for this form of care was driven by the hope that palliative care at the end of life would be cheaper than conventional, curative care:

> [During the 1970s] More and more talk was heard of 'cost containment' and 'the bottom line'. When the hospice movement knocked on the door, then, there was already a sense of urgency to discover some means of slowing the growth of health-related services. Since everybody dies, perhaps hospices could hold the costs down . . . (Mor et al., 1988: 15).

Hospice has flourished in the newly de-regulated, entrepreneurial environment of health care (Abel, 1986). As health care organizations have diversified their product lines to offer comprehensive cradle-to-grave services, hospices have sought to establish a "niche" in the continuum of care (Beresford, 1997). New funding streams help stimulate the growth of hospice programs (Laliberte & Mor, 1988). Many hospice organizations took advantage of the new funding made available through congressional legislation passed in 1982, which created the Medicare hospice benefit.[3] In 1989, congress increased reimbursement rates for Medicare-certified hospice care by 20% (NHO, 1999). While in 1984 there were only 31 Medicare-certified hospices[4] operating in the United States, by 1997 the number had grown to 2,274 (NHO, 1999).

THE PERCEIVED COST SAVINGS POTENTIAL OF HOSPICE HOME CARE

The three main features of hospice home care assumed to lower costs are: (1) the structure of Medicare hospice reimbursement, which functions to control services and costs; (2) the shift in the "locus of care" (Greer et al., 1988a: 236) from costly hospital care to cheaper home care; and (3) its reliance on informal caregiving labor, which shifts labor costs from the hospice agency to patients' families.

The Medicare Hospice Benefit as a Form of Managed Care

The Medicare hospice benefit – the federal program designed to provide terminally ill Medicare-eligible patients with a fixed package of hospice home care services – is a prospective payment system. This means that Medicare pays hospice agencies a fixed, daily rate for each of four categories of hospice patient services: (1) routine home care, (2) general inpatient care, (3) inpatient respite care, and (4) continuous home care. The rate is paid for each day the patient is under an agency's care, regardless of the agency's actual expenses. Medicare also imposes a "cap" on what they will pay per year, per patient. This reimbursement structure for hospice services forces agencies to take on the financial risk for each patient they accept, since the amount they can expect to be paid per patient is fixed. Thus agencies must continually balance how much they spend providing services with the amount they can expect to be reimbursed by Medicare.

Routine home care costs agencies the least to provide. The other three categories – general inpatient care, inpatient respite care, and continuous home care – require much higher financial outlays on the agency's part, whether they provide the services themselves (for example, by using their own registered nurses to provide continuous care) or contract out to other hospital and skilled nursing facilities. The most profitable patients, then, are those who: require the least intensive care; require only routine home care; and who have a strong informal caregiver network (Greer et al., 1988a). This reimbursement structure provides hospice agencies an incentive to strongly encourage patients to forego curative and costly inpatient treatment in favor of palliative home care only (Abel, 1986).

Shifting the locus of care

Perhaps the most important mechanism of cost savings in hospice home care is the shift in the location of care. Hospice home care saves money because it substitutes home care for inpatient care. In other words, it removes patients from costly hospital-based care and relocates them in the home, where care costs less (Mor & Kidder, 1985). For example, Hughes et al. (1992) found evidence that suggests the possibility of such a substitution effect. While the cost of home care for hospice patients in their study exceeded conventional care patients' home care costs, the net cost of hospice patients was much lower;

hospice patients used $1,639 less in hospital costs than conventional care patients – a reduction of 18%. In another study that compared hospice home care patients to conventional care patients, Kidder (1988) found that in the last month of life, home hospice patients' inpatient costs rose only 4% while conventional care patients' inpatient costs rose almost 250%.

Shifting Costs to Informal Carers

Hospice home care also limits labor costs by shifting a considerable portion of direct patient care onto patients and their families:

> If hospice care is cost-effective, it is because the family bears so much of the cost. Home care appears to be feasible only when the family 'pays' for the patient's presence at home by providing many hours of direct care. This cost shifting from the formal to the informal system is consistent with other recent health policies such as early hospital discharge and 'same-day surgery' (Greer et al., 1988a: 235–36).

Hospice organizations' ability to cut costs by substituting home care for inpatient care is directly related to this shift from formal to informal caregiving labor. Hospice patients require intensive, oftentimes round-the-clock care. If unpaid caregivers are not available, hospice agencies must cover the labor costs of formal health care workers. As Greer et al. (1988a) have noted, hospice agency revenues are directly related to the availability of unpaid caregivers; patients with strong informal caregiving support cost agencies less because they require fewer services. At the same time, they generate the same revenue as high-cost patients, many of whom have weaker informal caregiving support (ibid, 242).

Hospice's dependence on informal caregiving labor is part of a wider trend in healthcare policy which is increasingly emphasizing the role of informal caregiving labor (Benjamin, 1993; Glazer, 1993; Greer et al., 1988a). While there are no statistics available on the economic value of informal hospice caregiving in particular, Arno, Levine and Memmott (1999) have estimated that, during 1997 in the U.S., the value of all home health care provided by informal caregivers totaled $196 billion. They estimate the total value of formal home health care provided in the same year at $32 billion. Both home health care, generally, and hospice home care, in particular, rely heavily on the availability of unpaid caregivers to provide the vast bulk of patient care.

This increased dependence on informal caregivers represents an important change in the relationship between health care professionals and family members of patients. Health care professionals typically consider family members organizational "outsiders" and devise work practices that minimize their disruption of medical routines (Chambliss, 1996; Sudnow, 1967). As care is increasingly shifted from the hospital back to the home, this strategy becomes less available. Home care in general and hospice home care, in particular, fit the model described by Heimer and Staffen of organizations that are dependent upon people who fall outside the institutional system – when this is the case, they argue, members of the organization cannot afford to shut these people out. Instead, they must work to integrate them into the system in order to achieve the institution's goals (Heimer & Staffen, 1995). In hospice home care, nurses – who are dependent on informal caregivers to maintain patients in the home – must train caregivers into the values and goals of the hospice organization. Specifically, hospice home care nurses must train caregivers to adapt the hospice goal of supporting the patient's death at home.

SHIFTING NOTIONS OF THE GOOD DEATH

There is a growing literature on how hospice care has been transformed since its origins (Abel, 1986; McNamara et al., 1994; James & Field, 1992; Raymer, 1996; Beresford, 1997). Hospice, which sought to reform the over-medicalization of care of the dying, has become increasingly routinized and bureaucratized as it has sought incorporation into the mainstream healthcare system in the U.S. and other Western countries. Abel (1986) argues that the institutionalization of hospice has resulted in a comprehensive distortion of its original mission. In what follows, I briefly define the notion of the "good death" around which modern hospice is organized, focusing on patient autonomy as a central feature of the concept.

The Good Death Ideal

The concept of the "Good Death" recognizes that death, rather than simply a biological event, is a continuous social process experienced by a variety of individuals – the dying person, friends and family, and health professionals (McNamara et al., 1994). Awareness and acceptance of death, open communication, and involvement of the dying person's network of family and friends are seen as important components of a good death. Modern hospice is organized around this model. As Raymer writes, "hospice has always been about providing physical, emotional, social, and spiritual support for the terminally ill and their loved ones before, during, and after death" (1996: 53).

In their article tracing the development of the good death ideal, McNamara et al. (1994) suggest that the concept of the good death is not always "compatible" with the trend toward institutionalization, citing the basic tension that exists between the need for organizational stability and regularity on the one hand, and the desire to maintain hospice ideals on the other. Modern hospice continues to be organized around the goal of helping achieve a good death, but the meaning of the idea, they argue, is changing as hospice becomes more fully integrated into the mainstream system of healthcare. As James and Field note, "we see a shift in the thinking of those involved in hospice care from a broad concept of 'good' to a prescribed type of death, that is 'peaceful' " (1992: 1367). I argue in this paper that, in hospice home care, there is a further shift in the interpretation of the concept from "peaceful" to a death that occurs in the home. This shift is significant because it represents a further constricting of hospice's original holistic, patient-centered approach to care.

Patient Autonomy in Hospice

The notion of patient autonomy is central to the good death ideal. A good death is one in which the patient and caregiver actively participate in decision-making about the patient's end of life care. The increased emphasis on home death as the goal of hospice might not seem problematic, especially given that most people favor the idea of a home death. According to a 1996 National Hospice Organization/Gallup poll of Americans, about nine out of ten people said that they would prefer to be cared for in their own homes or in a family member's home if diagnosed with a terminal illness (NHO, 1999). But as Abel has noted, people who have chosen hospice home care sometimes reconsider entering the hospital as death approaches (1986: 79). Caregivers, too, may reconsider using hospital care when a patient's condition suddenly changes (Howarth & Willison, 1995) or when the cumulative stresses of caregiving become too much (Aneshensel, Pearlin & Schuler, 1993).

As organizational constraints have tended to restrict the goals of hospice, the value of patient autonomy in hospice has weakened. Nurses' attention to the needs of the organization helps limit and fix their working interpretations of what constitutes a good death (McNamara et al., 1994). As a result, the organization's needs for predictability can overwhelm the patient's desire to explore what the good death will mean for him or her. A patient/caregiver-defined good death might deviate significantly from the trajectory that a home hospice plan of care envisions. Patients' acceptance of death may vary from day to day and they may want to go in and out of the hospital. Or patients and their family members may wax and wane in terms of their desire for curative treatments. A home hospice course of care that includes any of the above will be interpreted by the organization's members as problematic – "bad deaths" (those which don't fit the normatively defined "peaceful," at home death) drain resources of money, time and organizational energy (ibid: 1506).

PREVENTING HOSPITALIZATION AT HHA

HHA receives patients via hospice referrals from physicians. The agency's decision to admit a patient to hospice care is made at the nurse's first "admit" visit. HHA guidelines define a "hospice appropriate" patient as someone who has a terminal diagnosis with a prognosis of 6 months or less to live, has decided to discontinue curative treatments, and wants to die at home. On this last criterion – the desire to die at home – patients are not always clearly decided. HHA nurses understand that part of their job on the admit visit is to discern what patients and their potential caregivers are likely to choose (staying at home versus opting for hospital care) as the disease progresses. There is no explicit rule that patients cannot enter the hospital. In practice, however, the agency employs a fairly rigid informal rule regarding hospitalization. For HHA nurses, if a patient goes to the hospital, the event is understood as, in the words of one nurse, a "break in really what it is that we do."

There are two reasons to believe that preventing hospitalization is an important issue for HHA staff. First, in case conferences (where newly admitted and ongoing patients are discussed) the interdisciplinary team pays a lot of attention to the issue of who is likely to seek inpatient care. Second, when a patient enters the hospital without first revoking the hospice benefit, especially if it is not one of two hospitals with which HHA has contracted for limited inpatient hospice care, HHA's policy requires that the patient's hospice benefit be revoked. I observed this happening on several occasions when the nursing staff found on the triage report in the morning (the document that lists hospice activity during the previous night) that a patient had been admitted to the hospital during the night without having contacted the agency first. In these instances, the supervisor contacted the patient's family to ask that they sign a "Revocation of the Hospice Benefit" form. If the patient's family refused to revoke the benefit, the supervisor submitted an unsigned revocation form and documented that the disqualified patient (or patient "surrogate") had refused to sign the revocation.[5]

The agency's concerns about keeping patients out of the hospital are evident during all phases of a patient's hospice career. Decisions about whom to admit as a hospice patient take into account patients' and caregivers' attitudes about hospitalization versus home care. During the course of ongoing care, nurses work to sustain the caregiver's desire and ability to provide home care. And finally, nurses become especially attuned to the issue of preventing hospitalization as patients enter the active dying stage. While there are individual variations in how the nurses I studied deal with this issue, there are several strategies HHA nurses commonly use to minimize the possibility of hospitalization. In what follows, I describe three of the strategies nurses employ to discourage hospitalization.

1. Assessing Patient Appropriateness and Caregiver Competence. Admitting decisions at HHA involve assessments about whether patients are "appropriate" for hospice and about caregivers' potential competence to provide care. A terminal diagnosis does not in itself determine whether a patient is hospice appropriate. How the patient and caregiver will react to the diagnosis must be taken into account since their actions may cost the agency more than expected. Thus the patient's and caregiver's concrete orientation to terminality and hospice care are pivotal factors in determining hospice appropriateness. The agency's financial concerns create the "organizational conditioning of interpretation" (Gubrium & Holstein, 1997: 207) which influences the hospice team's admitting decisions.

Nurses' assessments of caregivers' potential competence include attention to this issue. In addition to deciding whether a caregiver will be able to provide care, nurses assess whether or not caregivers are likely to follow through on the patient's hospice plan of care, which assumes that the patient will die at home. Even in circumstances where the patient is clearly going to die soon and could benefit from services covered by the hospice benefit, the caregiver's orientation to the goal of a home death affects decisions to admit, as in the case of this patient – a young woman with late-stage cancer who weighed only 70 pounds and whose kidneys were actively failing:

> [The patient's sister was saying] "You're just gonna let her die!" as if that was our choice . . . they talked about the patient not being able to swallow and that they wanted to put a tube down her to force feed her if that happened . . . this was a case where there was no money and I could not certify [the patient] for hospice . . . because [the caregivers] wanted the option of taking her to the hospital for aggressive care, which made her really not appropriate for a certified program (interview).

When the nurse feels sure that the caregiver will insist on hospitalization or other "inappropriate" treatments, as in this patient's case, the caregiver's refusal to accept the goals of palliative care and home death becomes the overriding factor in the decision not to admit the patient under the hospice benefit.

Some (not all) patients like the one described above are admitted using HHA's informal system of various hospice "statuses." This system of quasi-hospice categories ("pre-hospice," "regular hospice," and "county bill") allows the nurses to provide different levels and types of hospice care without committing the agency to the risk sharing relationship built into the hospice benefit. In other words, for patients who, for various reasons, are not considered hospice appropriate, HHA provides quasi-hospice services while being reimbursed through either regular Medicare payments, MediCal, or the county program for indigent patients. The important point here is that this informal system allows HHA to categorize different types of patients based on whether they are likely to use hospital care, while at the same time limiting the agency's financial risk by certifying only the patients at low risk for hospitalization.[6]

Assessing the caregiver's competence is not something the nurse does once and for all, however. Nurses' judgments change and develop over time as the patient approaches death and as both the nurse and the caregiver negotiate ongoing care. Nevertheless, nurses pay close attention to this issue in their initial assessments of patients' caregiving situations.

2. Mobilizing Emotional Connection and Caregiver Commitment. Nurses view emotional connection to the patient as an important part of what motivates the caregiver and creates a sustained commitment to hospice care. Nurses frequently invoke this aspect of the caregiver/patient relationship when assessing what they might be able to expect from caregivers. In case conference, families who are described as "close," "together," "devoted," or in whose homes can be found evidence that "there have been some good times," are typically expected to have the emotional resources that help create sustained commitment to supporting the patient's death at home.

Nurses actively cultivate the emotional connections between caregiver and patient in order to build commitment to the idea of home death. By focusing on the caregiver's relationship to the patient, the nurse tries to both invoke and confirm a compelling sense of responsibility to help the patient fulfill his or her wish to die at home. Even where there is no family or kin tie, the same approach can help reassure paid caregivers of the importance of their commitment to the patient.

This strategy of working with emotional connection was evident in the case of John Roll, a patient whose primary caregiver (his wife) was judged by the nurse to be unable to provide care because she displayed symptoms of dementia. On our way to the Roll's home, the nurse explained that she was preparing to convince the family to hire a caregiver, even though they had shown some reluctance to do so in her previous visit. Mr. Roll's daughter, whom the nurse had not met before, was there along with the patient's wife when we arrived. During the visit, the nurse engaged the daughter in matters about her father's care and condition, testing her grasp of the situation and the caregiving regime. She sensed that the daughter's emotional connection was an important part of her involvement with her father – the daughter had only just arrived from out of town the day before, yet the nurse was impressed with the quality of her interaction with her father and her grasp of his needs:

> I could see that things were going to work out . . . based on how the daughter was talking with me, that she was committed . . . that she wanted to be the one to be the caregiver. She wanted to be involved (interview).

The daughter needed to return home after the weekend because of work commitments, but her father's physical signs indicated he was "actively dying." The nurse needed to convince the daughter to take over the role of primary caregiver now as opposed to next week, as she had originally planned. In what follows, the nurse appeals to the daughter's emotional connection to both her father (the patient) and her mother:

> The nurse listens to the daughter talk about last night and then says, "it's really hard for you, isn't it?" She praises her for doing such a good job and says, "no one could do this like you're doing it." The daughter, crying, says, "I know, I'm glad I could be here but it's really hard." The nurse just looks at her nodding her head and tells the daughter she's really glad to learn that she's here and thinks she ought to stay because . . ." your mother just can't handle this." After a long pause, the daughter agrees to stay and makes a call to change her work schedule (patient visit).

This example highlights the nurse's active management of emotional connection in order to recruit a caregiver capable of facilitating the patient's home death.

In other instances, nurses invoke the caregiver's connection to the patient in the course of routine care, especially at times when they sense that a caregiver may be having doubts about her ability to attend to the home death, as in the following example of Helen Lake, a 102 year old patient who lived in a retirement condominium. In this case, the nurse had just finished discussing with the paid primary caregiver what to do when the patient dies. The caregiver was clearly uncomfortable with the prospect of being alone when it happened:

> The nurse tells the caregiver that when the patient stops breathing not to panic and to take her time. "It's O. K. to leave and go downstairs to tell someone if you need to," she tells her. The nurse then looks at the caregiver warmly and says, "you've been with her a long time now, huh?" The caregiver nods and they talk about how long it's been and what a nice lady the patient is (patient visit).

This example is characteristic of the way that nurses highlight and refresh the caregiver's connection to the patient as an ongoing part of trying to sustain the caregiver's commitment to the hospice plan of care.

3. Normalizing Death at Home. If caregivers are going to remain committed to the patient's home death, the nurse must be concerned with managing caregivers' anxiety about attending to a home death. HHA nurses try to alleviate the caregiver's anxiety by helping the caregiver interpret the patient's death at home as an expected, and thus normal, event. In the following excerpt, the nurse works with the family/caregivers of a patient who is very near death to reframe how they think about what is happening:

> [Elderly patient – comatose, septic, very near death] – the nurse . . . moves into the kitchen to talk with members of the patient's family and her friend. She asks the patient's friend and the son, "Do you know what to do when she dies?" The friend says, "Call 911." The nurse responds quickly, "no," and explains to them both that they should call the hospice when she dies. "You want her to die at home, right?" She asks and the friend responds, "yeah, that's why we brought her home from the hospital." But the son wants to know if she'll be given antibiotics for her infection. "No, that's not appropriate for her, we just need to keep her comfortable" the nurse answers. She continues instructing them to call hospice when she dies (patient visit).

In my later interview with this nurse, she reported having been worried that she might have to revoke the patient because the family/caregivers seemed likely to call 911.

Nurses' reframing of the death at home as normal highlights their dependence on informal caregivers: whereas family members are typically shut out of healthcare practices and routines, in the case of hospice home care, health professionals find themselves having to train family members into the routines of the health care system. In other words, HHA nurses must try to impart to family members a sense of the routineness of home death. Being able to routinize the "disaster" of their loved one's death is part of what makes it possible to achieve the emotional distance necessary to manage the emotional and physical intensity of the caregiving task (Chambliss, 1996). Thus HHA nurses work with caregivers on an ongoing basis to train them to understand that home deaths are common, achievable, and preferable to hospital deaths.

The way that HHA nurses deal with peripheral family members points towards the importance of helping caregivers reframe the home death as normal. Oftentimes the cast of family members involved in a patient's care changes over time, with people dropping out or coming on the scene late in the patient's course of home care. Caregivers who have been involved in patient care over time and in concrete ways have had the opportunity to adapt to the idea that things are progressing as they should. In contrast, family members who have not been involved in the early stages of the patient's care sometimes arrive toward the end and react in disruptive ways, a phenomenon one nurse calls "out of town syndrome." Family members who have not had a chance to adopt a commitment to the idea of the patient's home death often panic toward the end, disrupting the course of care:

> When you have a family on service and everything's going really smoothly but then you have relatives coming from out of town that are coming in to visit . . . you can almost guarantee that there's gonna be someone in the family that, all of a sudden, wants

> everything done . . . "why isn't this patient on IVs!?."..and when the patient expires they want to call 911 (interview).

Anyone who is only peripherally involved but who has the power to influence decision making can present this problem because they can easily be overcome with the sense that something more should be done. The nursing clinical supervisor, who oversees HHA office policy and practices, deals with this problem by encouraging the nurses to involve peripheral family members by giving them caregiving tasks. Giving family members something physical to do, she feels, gives them a concrete sense of participation and control over the process of care:

> [When she was assigning a new patient to one of the nurses] I said to her . . . I would involve her sister AS MUCH as possible, you know. Teach her to do all this stuff, too [wound care, medications pump, etc.] because that way, she will be so involved it'll give her the feeling like she has some participation (interview).

The supervisor understands that when caregivers do hands-on care, they learn to accept what is happening as routine and thus are more committed to supporting the patient's death at home.

While nurses work throughout the course of the patient's care to reframe the death at home as normal, they become especially attentive to this issue when the patient enters the "active dying" stage, the time frame usually within one week to two days of the patient's death. The physical changes that come with active dying can threaten the caregiver's ability to view home death as a routine event. To try to prevent caregivers from panicking, nurses try to anticipate for caregivers what might happen, as one nurse explains:

> Generally it'll come up [hospitalizing the patient] for people who suddenly become short of breath or if there's a sudden, overwhelming symptom that they're not prepared for . . . they feel they need to do something about it . . . if someone has a seizure that's pretty scary to watch and feel like not calling 911 about unless the family's been well prepared in advance (interview).

Dramatic events like these can overwhelm caregivers with anxiety, thus making them more likely to seek emergency, inpatient care. Under these circumstances even caregivers who have been able to achieve a somewhat distanced sense of the routineness of what is happening can change their minds about wanting the patient to die at home. Even in situations that most non-medically trained people would find extremely daunting, HHA nurses try to encourage caregivers to think of themselves as capable of managing the death at home, as the following excerpt illustrates. In it, the nurse explains the situation of one of her patients who has a massive tumor in his abdomen:

> His tumor's growing . . . he's the type of person, he's gonna have a sudden death. And I've warned them. I've gone over with them what to do because my suspicion is that he's gonna

> bleed out suddenly because it's wrapped around the aorta. So it's just gonna eat through . . . [He'll] either start vomiting blood or bleeding or something . . . They're prepared. They know. And I told them, you know, just keep dark towels around and just, you know, put pressure on it and to stay with him and call me and I'll see if I, if it's at night or something, I live around the corner from them, so I told them I would come (interview).

Thus two important and interrelated strategies nurses use as the patient enters the active phase of dying are to prepare the caregiver concretely for how to handle what may come and to draw the caregiver into a frame of mind that will encourage his or her continued commitment to the goal of home death.

HHA nurses actively try to prepare caregivers for any eventuality for two reasons. First, they are explicitly expected by the organization to work toward this goal. Second, they are also personally inclined to believe that dying at home is best and that they can effectively train caregivers to help patients achieve that goal. But they balance their convictions that death at home is possible for all patients with an understanding that, in the real world, some people cannot manage it. As one nurse confided to me, "caring for someone 24 hours a day is really not a human thing to do." When patients and/or caregivers decide they want the support of inpatient hospital services, HHA's practical policies require that the patient's hospice benefit be revoked. Thus while the hospice benefit is designed to allow some limited access to inpatient care (through the general inpatient and inpatient respite provisions), home hospice care at HHA tends to be an all or nothing proposition.

THE TRANSFORMATION OF HOSPICE CARE

I have described several ways in which HHA nurses control home hospice caregivers' abilities to support terminally ill patients' death at home: (1) through their assessments of patient and caregiver appropriateness and competence; (2) by appealing to the caregiver's emotional connection to the patient in order to encourage commitment to the goal of death at home and; (3) by normalizing the idea of death at home. By managing caregivers' abilities to maintain dying patients in the home, hospice nurses provide an important new mechanism for dehospitalizing dying patients.

I see this increased emphasis on dehospitalization in hospice work as part of what Abel calls "the transformation of hospice care from a human service into a commodity" (Abel, 1986: 82). This transformation involves standardizing the goals and practices of care in order to reduce costs. One way of achieving this end is by narrowing the scope of possible outcomes of care. At HHA this means organizing care to reduce the likelihood that patients will request and use inpatient hospital care. The work practices used by HHA nurses exert a subtle influence on patients and caregivers – they shape caregivers' understandings of both their own responsibilities and what they can reasonably expect from the health care system.

HHA nurses are motivated by deep personal and ideological commitments to patient-centered care. At the same time, as I have shown, they are influenced by the organizational context within which they work. Nurses' notions of what constitutes the good death are shaped by the agency's objective of providing home-based care. As a result, the broad concept of the good death that originally guided hospice work is gradually being replaced by the narrower goal of having death occur at home.

In most instances this goal of home death is shared by patients and caregivers. Nevertheless, there remains a constant tension around this issue in the work that HHA nurses do. If patients "die as they lived," as the popular hospice saying goes, then we might expect that hospice care should make room for variability in how people feel about their impending deaths and about what courses of action they will choose to take. This tension arises when patients and caregivers sometimes struggle to exercise more control and flexibility over the process of hospice care than HHA's organized work practices allow. For example, some patients and/or their caregivers are comfortable with home care until the final phases of terminal illness, at which time they feel more comfortable with hospital care. Instances where caregivers call 911 in the night often reflect the caregiver's desire for a broader range of treatment possibilities than the hospice plan of care allows. Nurses' supervision of hospice caregivers thus restricts patients' and caregiver's access to a full range of end of life care options.

This transformation of hospice home care is closely related to changes in the labor process currently affecting the health care delivery system (Glazer, 1993; Gordon, 1997). In the case of seriously or terminally ill patients, when inpatient care is limited, the need for accountable, direct care remains. Thus when health care organizations limit utilization in order to cut costs, they become increasingly dependent on the informal sector to provide appropriate care for their patients (Houts, Nezu, Maguth Nezu & Bucher, 1996). This is not simply a matter of "coercing" informal caregivers to do the work in a "new division of labor" as Glazer argues (1993: 6). Rather, it is best understood as an institutionally-conditioned process in which health care professionals, in order to meet the demands of their employing organizations, must actively recruit, assess, train, and motivate these informal caregivers to work cooperatively with the health care system. Through their work of training caregivers to equate the good death with a home death, HHA's fiscal and management concerns penetrate caregivers' and patients' notions about what constitutes a good death and appropriate end of life care and, as a result, their use of services.

CONCLUSION

The growth of hospice as a commodity has important implications for issues of access and quality of care. As hospice becomes increasingly concerned with preventing patients' hospitalization, it is likely that access to hospice care for patients from traditionally underserved groups (e.g. racial and ethnic minorities, low-income people, socially isolated and otherwise difficult to serve patients) will worsen and the range of patients who meet eligibility criteria will narrow. As Brenner notes (1997), there is ample evidence that hospice care is disproportionately provided to white, middle class patients. To the extent that there are ethnic or other group variations regarding attitudes toward hospice and palliative care, access to insurance coverage, or families' ability to provide care, the already existing gap in access to hospice services is likely to grow. Moreover, as the logic of cost-savings cuts further into hospice care, causing nurses to increasingly straddle the roles of patient advocate and utilization manager, there is likely to be a further decrease in patient autonomy in decision-making about end of life care, one of the hallmarks of hospice care.

Given projections regarding the aging of the population and hospice's growing importance in the long-term care continuum (Greer, Mor & Kastenbaum, 1988b), I argue that the care delivery practices I have described here will play an increasingly important role in changing societal values regarding the appropriate use of healthcare resources at the end of life. For this reason, more research is needed that can aid our very limited understanding of how hospice home care services are actually delivered and used, and how both interactional and organizational factors affect service provision, utilization and outcomes.

NOTES

1. The agency's name and names of patients are pseudonyms.
2. These figures are for hospice care in general, which includes hospital – and skilled nursing facility-based hospice, freestanding inpatient hospice and home-based hospice care. The majority of all hospice care in the U.S. is delivered in the home setting.
3. See Mor, Greer and Kastenbaum (1988: 12) and Greer, Mor and Kastenbaum (1988a: 239) for details regarding Medicare hospice program eligibility requirements and coverage.
4. Medicare was the primary payment source for 66.8% of all hospice care delivered in the U.S. in 1994. Medicaid covered 9.1%. Private insurers were responsible for coverage of 14.6% of hospice care in the same year and programs serving the indigent paid for 6.3%, while the remaining 3.3% was covered by other sources (NHO, 1999). In my discussions of reimbursement for hospice services, I focus exclusively on the

Medicare hospice benefit since it is the dominant payment source for hospice home care.

5. When patients switch their coverage from the hospice benefit back to regular Medicare their medications costs are no longer fully covered. They also lose continuity of care in many cases because of a lack of adequate coordination between hospital and hospice. HHA actively limits the extent to which they will use contracted hospital services for their patients, even though the hospice benefit is designed to allow limited use of inpatient services. One reason for this is that the agency's costs increase. Nurses also limit these arrangements because they lack confidence in hospital staff's ability to provide palliative care and also because such arrangements greatly complicate nurses' efforts to coordinate their patients' care.

6. HHA's system of hospice status categories was dropped toward the end of the year I was in the field after the organization was bought out by first one and then another large healthcare corporation.

ACKNOWLEDGMENTS

I thank Natasha Chen Christensen, Tiffani Chin, Amy Dennisen, Tim Diamond, Bob Emerson, John Heritage, Katherine Hill, Maggie Kusenbach, Christine Morton, Judy Seltzer, and the editor for their helpful comments on this and previous versions. This investigation was supported by National Research Services award T32 HS00046 from the Agency for Healthcare Research Quality.

REFERENCES

Abel, E. K. (1986). The Hospice Movement: Institutionalizing Innovation. *International Journal of Health Services*, *16*(1), 71–85.

Aneshensel, C. S., Pearlin, L. I., & Schuler, R. H. (1993). Stress, Role Captivity, and the Cessation of Caregiving. *Journal of Health and Social Behavior*, *34*, 54–70.

Arno, P. S., Levine, C., & Memmott, M. M. (1999). The Economic Value of Informal Caregiving. *Health Affairs*, *18*(2), 182–188.

Benjamin, A. E. (1993). An Historical Perspective on Home Care Policy. *The Milbank Quarterly*, *71*(1), 129–166.

Beresford, L. (1997). The Future of Hospice in a Reformed American Health Care System: What Are the Real Questions? *The Hospice Journal*, *12*(2), 85–91.

Brenner, P. R. (1997). Issues of Access in a Diverse Society. In: B. Jennings (Ed.), *Ethics in Hospice Care: Challenges to Hospice Values in a Changing Health Care Environment* (pp. 9–16). New York: Howarth Press.

Brooks, C. H., & Smyth-Staruch, K. (1984). Hospice Home Care Cost Savings to Third-Party Insurers. *Medical Care*, *22*(8), 691–703.

Chambliss, D. F. (1996). *Beyond Caring: Hospitals, Nurses, and the Social Organization of Ethics.* Chicago, IL: University of Chicago Press.

Dee-Kelly, P. A., Heller, S., & Sibley, M. (1994). Managed Care: An Opportunity for Home Care Agencies. *Nursing Clinics of North America*, *29*(3), 471–481.

Glazer, N. Y. (1993). *Women's Paid and Unpaid Labor: The Work Transfer in Health Care and Retailing*. Philadelphia: Temple University Press.

Gordon, S. (1997). The Impact of Managed Care on Female Caregivers in the Hospital and Home. *Journal of the American Medical Women's Association*, *52*(2), 75–77, 80.

Goodlin, S. J. (1997). Hospice Organizations' Role in Health Care Improvement. In: B. Jennings (Ed.), *Ethics in Hospice Care: Challenges to Hospice Values in a Changing Health Care Environment* (pp. 71–80). New York: Howarth Press.

Greer, D. S., Mor, V., & Kastenbaum, R. (1988a). Public Policy and the Hospice Movement. In: V. Mor, D. S. Greer, & R. Kastenbaum (Eds), *The Hospice Experiment* (pp. 227–244). Baltimore, MD: Johns Hopkins University Press.

Greer, D. S., Mor, V., & Kastenbaum, R. (1988b). Concepts, Questions, and Research Priorities. In: V. Mor, D. S. Greer & R. Kastenbaum (Eds), *The Hospice Experiment* (pp. 245–251). Baltimore, MD: Johns Hopkins University Press.

Gubrium, J. F., & Holstein, J. A. (1997). *The New Language of Qualitative Method*. New York: Oxford University Press.

Heimer, C. A., & Staffen, L. R. (1995). Interdependence and Reintegrative Social Control: Labeling and Reforming 'Inappropriate' Parents in Neonatal Intensive Care Units. *American Sociological Review*, *60*, 635–654.

Houts, P. S., Nezu, A. M., Nezu, C. M., & Bucher, J. A. (1996). The Prepared Family Caregiver: A Problem-Solving Approach to Family Caregiver Education. *Patient Education and Counseling*, *27*, 63–73.

Howarth, G., & Willison, K. B. (1995). Preventing Crises in Palliative Care in the Home: Role of Family Physicians and Nurses. *Canadian Family Physician*, *41*, 439–445.

Hughes, S. L., Cummings, J., Weaver, F., Manheim, L., Braun, B., & Conrad, K. (1992). A Randomized Trial of the Cost Effectiveness of VA Hospital-Based Home Care for the Terminally Ill. *Health Services Research*, *26*(6), 801–817.

Hughes, S. L., Ulasevich, A., Weaver, F. M., Henderson, W., Manheim, L., Kubal, J. D., & Bonarigo, F. (1997). Impact of Home Care on Hospital Days: A Meta Analysis. *Health Services Research*, *32*(4), 415–432.

James, N., & Field, D. (1992). The Routinization of Hospice: Charisma and Bureaucratization. *Social Science and Medicine*, *34*(12), 1363–1375.

Kidder, D. (1988). The Impact of Hospice on the Health-Care Costs of Terminal Cancer Patients. In: V. Mor, D. S. Greer & R. Kastenbaum (Eds), *The Hospice Experiment* (pp. 48–68). Baltimore, MD: Johns Hopkins University Press.

Laliberte, L., & Mor, V. (1988). Organizational Growth and Change Among Hospices. In: V. Mor, D. S. Greer & R. Kastenbaum (Eds), *The Hospice Experiment* (pp. 209–226). Baltimore, MD: Johns Hopkins University Press.

McNamara, B., Waddell, C., & Colvin, M. (1994). The Institutionalization of the Good Death. *Social Science and Medicine*, *39*(11), 1501–1508.

Mahoney, J. J. (1997). Hospice and Managed Care. *The Hospice Journal*, *12*(2), 81–84.

Mor, V., Greer, D. S., & Kastenbaum, R. (1988). The Hospice Experiment: An Alternative in Terminal Care. In: V. Mor, D. S. Greer, & R. Kastenbaum (Eds), *The Hospice Experiment* (pp. 1–15). Baltimore, MD: Johns Hopkins University Press.

Mor, V., & Kidder, D. (1985). Cost Savings in Hospice: Final Results of the National Hospice Study. *Health Services Research*, *20*(4), 407–422.

National Hospice Organization (1999). Basic Statistics About Hospice, online fact sheet: www.nahc.org/Consumer/hpcstats (March 29, 1999).

Raymer, M. (1996). Revisiting the Core Values of Hospice. *CARING Magazine*, November 1996, 52–55.

Schatzman, L., & Strauss, A. L. (1973). *Field Research: Strategies for a Natural Sociology.* New Jersey: Prentice-Hall.

Sudnow, D. (1967). *Passing On: The Social Organization of Dying.* New Jersey: Prentice-Hall.

CAPITAL FORMATION AND THE UNITED STATES HEALTH CARE SYSTEM: THE RELATIONSHIP BETWEEN THE PRIVATE AND THE PUBLIC SECTOR[1]

Beth Mintz and Michael Schwartz

ABSTRACT

In this chapter, we begin to clarify the role played by corporate capital in the development of the United States health care system. To do so we explore the role of capital, more generally, in health care expansion and we broaden the traditional focus on capital within the private sector to include public expenditures and subsidies as part of a system of capital flows that constrains corporate development. We consider government subsidies and health insurance reimbursement practices as capital formation mechanisms and we investigate the history of health care funding in this context. We understand the development of the U.S. health care system as driven by the interaction between the public and private sectors: Actions of the state create constraints on private investment and subsequent private investment decisions create equally compelling constraints on future public policy. We view the rise of corporate capital as a constrained consequence of these earlier dynamics. Moreover, we suggest that government spending patterns can have as profound an effect

Health, Illness, and Use of Care: The Impact of Social Factors, Volume 18, pages 229–248.

ISBN: 0-7623-0740-4

on corporate development as legislative activity and we believe this is has important implications for the development of health care systems.

INTRODUCTION

The United States health care system continues to undergo profound change and the introduction of corporate capital into health care delivery has been particularly important in this transformation. In recent years, it has joined government funding and private insurance reimbursement as vehicles of health care financing, although it remains the least understood of the three. We are interested in clarifying corporate capital's role in health care expansion but we believe that to consider this, we first must understand the larger role of capital in the development of the modern health care system. Thus, in this work we explore capital's role in health care development and then trace the rise of corporate capital, more specifically, as a constrained consequence of earlier dynamics.

In sociology, most discussions of capital and capital formation concentrate on the private sector as a major source of funding. Commercial banks and insurance companies, for example, have been studied in detail and some suggest that control over capital flows concentrates broad decision-making power in the hands of financial institutions (Glasberg & Schwartz, 1983; Mintz & Schwartz, 1985). Here, we broaden this focus on capital to include public expenditures and subsidies, which we believe can be scrutinized as part of a system of capital flows that constrains corporate development. We suggest that government spending patterns can have as profound an effect on corporate development as legislative activity and we believe this is true for the development of health care systems, as well.

Moreover, we see the relationship between corporate capital and the state as interactive. Actions of the state constrain private investment and subsequent profiles of private investment create equally compelling constraints on future public policy. We understand the development of the American health care system, then, to be a product of the interaction between business and state decision making and we suggest that decisions over the nature of capital flows ultimately determined the trajectories of health care as an industry and as national public policy.

CAPITAL AND HEALTH CARE

The concept of capital is generally applied to the private sector and most agree that the modern industrial enterprise needs large sums of capital for operating expenses, expansion, interest repayment, etc. Even though health care has traditionally been dominated by nonprofit organizations and that health care markets are noncompetitive in a formal sense (Ginsberg, 1998), the modern health care organization is more and more subject to the same capital requirements as private sector concerns. Understanding capital as any asset that can be used to generate income in both the narrow sense of funding everyday operating expenses and in the broader sense of fueling replacement or expansion of plants and equipment (Munn et al., 1991), we believe that this view of capital and the corresponding concept of capital formation applies quite well to the health care case.

Investigations of the development of the United States health care system have explored the role of both equity capital and debt financing. A third source of capital, retained earnings, has not generally been considered in this context, however, although in the broader literature the ability to generate capital within the firm has been viewed as increasingly important in understanding the modern corporation (Galbraith, 1967; Berle & Means, 1968; Penrose, 1995). Any policy that increases the retention of earnings, then, increases the ability to form capital. And with increased capital supply companies have the resources to expand, either in reaction to environmental pressure or as an industry innovator.

We believe that this view of retained earnings as a source of capital is extremely useful in helping us understand the development of the modern health care system. We also believe that the ability to generate funds internally is augmented by state policies and this is particularly relevant here. At least since World War II, governmental funding and both governmental and nongovernmental health insurance plans have been major actors in the financing of United States health delivery. While neither government subsidies nor health insurance reimbursement practices have traditionally been viewed as capital formation mechanisms, in the U.S. case they have operated in precisely that way and, therefore, have been crucial in determining the direction of growth of the health care system.

In this study, we investigate the role of traditional health care funding mechanisms in the development of the modern health care system. Our portrait of the history of U.S. health care is not new; the forces most instrumental have been discussed numerous times (c.f. Starr, 1982; Heirich, 1998) with many different emphases (c.f. Light, 1996). Here, however, we understand that development in a different way and we do so by considering capital as the liquid and fixed assets available to operate a business, even a nonprofit one.

We begin by distinguishing between debt financing and equity capital. For the former, insurance companies become the central focus of our analysis. For the latter, we concentrate on the entry of the private sector in health care delivery suggesting that their role in the system is a function of the already established dynamics of the earlier system. We, therefore, begin with a discussion of the logic of the insurance industry as a major lender and then turn to a review of traditional health care funding mechanisms. Finally, we trace the rise of corporate capital as an outcome conditioned by earlier decision making.

THE PRIVATE INSURANCE INDUSTRY

When the topic of insurance is raised in the context of health care, many of us immediately think of the role of reimbursement patterns in the development of health delivery systems. For at least 100 years, however, insurance companies have dominated long-term capital markets within United States (Mintz & Schwartz, 1985) and, here, we emphasize its importance as a source of capital. We suggest that an understanding of the role of the insurance industry as a capital supplier is central to an analysis of both the development of the U.S. health care system and to our understanding of capital formation within the sector.

In recent years, life insurance companies have held about $11 trillion in policy obligations (Kopcke, 1995) and the logic of these transactions is straightforward: they collect premiums from their clients that they invest in capital markets, long-term bonds most typically.[2] The profitability of the enterprise turns on both the number of clients insured and the rates of return on their investment portfolios. As a result, insurance companies maintain enormous bond holdings in major companies and the importance of this is highlighted when we remember that most long-term corporate debt is in bonds, rather than stocks. In 1995, for example, the life insurance industry held over $1.2 trillion in corporate America (Moody's Bank and Finance Manual, 1998). In the early 1990s, they held about one third of all corporate bonds issued in the U.S. – including for-profit health care concerns – and supplied a net total of about 30% of all funds raised by bonds within a 15 year period (Kopcke, 1995).

Traditional health insurance is driven by a similar dynamic, but the increasingly short duration of health insurance holdings – policy payouts are frequent – make them a poor source of investment capital. In 1988, for example, health insurance accounted for almost 25% of premium income, but only about 15% of profits (Weiss, 1992). And as health care costs spiraled, investment income was not able to shield the health insurance industry from market forces: in 1989, the 12 largest health insurers lost $830 million (Bodenheimer, 1990).

This understanding of health insurance logic makes the behavior of the industry more comprehensible. Many view third-party, fee-for-service, reimbursement patterns as a central mechanism behind health care inflation (Starr, 1982), and this is crucial for understanding the rapid growth of managed care, two points that we will address below. Here we note that higher costs meant larger pools of investment capital, attractive even for short-term use. Thus since industry profits turn on investment returns rather than premiums, there was little incentive to control costs, especially if they could be passed on to policyholders (Weiss, 1992). But when institutional purchasers rebelled against runaway costs and forced a change in the price setting mechanism, health insurance became a less attractive industry segment.

Ironically, as health insurance became less and less attractive, the insurance industry lost its market share to a new institutional adaptation: self-insurance. This was facilitated by government legislation exempting self-insured corporations from state regulations and from state insurance taxes and assessments. By 1988, 66% of medium and large corporations had turned to self-insurance (Weiss, 1992), thus making the employer the principal bearer of health care financial risks (Health Insurance Association of America, 1990).

As health insurance became a less and less reliable source of investment income, the health insurance industry moved into managed care, particularly into the vertically integrated networks of HMOs, hospitals, and physicians that have become so popular. Note that it is the resources of the insurance industry in general, collected from individual policyholders and invested in corporate stocks and especially bonds, rather than the financial power of health insurance specifically, that has fueled this expansion. And it is this enormous asset base that has led to the consolidation of the managed care into an industry dominated by a handful of insurance companies.

The insurance industry, then, has played a dual role in the development of the United States health care system. As an insurer, it has been an important actor in health care inflation and this will be discussed in detail below. As a major source of investment capital, it has encouraged the growth of specific organizational forms, it has played an important role in the development of private multi-hospital systems, and it has encouraged for-profit health care, in general. Currently, its asset base continues to help integrate insurance and provision, thus, changing the complexion of both health insurance and health delivery.

HEALTH CARE FINANCING: A BRIEF HISTORY

For the latter part of the twentieth century, government funding and nongovernmental health insurance plans have provided the overwhelming majority of capital for the health care delivery system. Health insurance is not overtly a capital formation mechanism or a vehicle for the expansion of health care. Traditionally, it was organized on a fee-for-service basis in which hospitals were reimbursed for the costs of services rendered and, with the introduction of Medicaid and Medicare, ultimately including equipment, depreciation and the costs of debt servicing. But this open-ended commitment to reimburse for all costs created an incentive to develop and implement high-priced procedures and technologies, since these would be inevitably reimbursed once they were brought on line. Starr, for example has argued that this system has been the central mechanism of health care inflation (Starr, 1982). Originally developed by Blue Cross, with strong input from the hospital industry, this is the model that informed later government health insurance programs. As fee-for-service insurance plans grew over time providing larger pools of capital, state entry into health care financing increased expansion possibilities greatly and supplemented the private sector's contribution to health care expansion.

Despite the built-in incentive for new construction and embellished treatment plans, fee-for-service reimbursement, even with its generous depreciation schedules, could not generate the up-front costs of hospital construction and technological development. In the early stages, capital formation was more difficult still. Prior to the Great Depression, the actual construction of hospitals echoed the great divide between public and private hospitals: federal, state and local governments financed the construction of public hospitals, while private hospital construction and expansion was dependent on philanthropy. It was not until after World War II that the government encouraged private (non-profit) hospital development, using the Hill-Burton Hospital Survey and Construction Act of 1946, as the vehicle. Commonly referred to as the Hill-Burton Act, this legislation marked the beginning of governmental attempts to solve private capital shortage problems in the health care sector.

Three things stand out about Hill-Burton funding. First the combination of the Hill-Burton Act and Blue Cross – the tax-exempt employee health plan mentioned above – provided expansionary capital which guaranteed the rapid growth of the medical industry (Himmelstein & Woolhandler, 1988). Whereas Blue Cross provided after-the-fact reimbursement, which encouraged expansion, Hill-Burton, solved the problem of initial capital formation by providing construction costs for project after project. Between 1947 and 1971, \$12.8 billion dollars in federal, states and local funds were generated for hospital construction (Starr, 1982). New hospital construction in the 1950s and 1960s accounted for about 40% of the inpatient capacity of the 1980s and by then, high tech equipment – also financed by Hill-Burton – became commonplace (Himmelstein & Woolhandler, 1988).

Second, the Hill-Burton Act did not provide for active state monitoring of the distribution and utilization of program funds (Starr, 1982). This was the beginning of a long-term strategy of governmental financial support without concurrent decision-making authority on specific policies. Note that this echoes private (and nonprofit) sector health purchases: it was not until the 1980s that (nongovernmental) institutional purchasers of health care began demanding input into health decision making, although in the 1970s, Congress began a series of attempts at controlling costs through regulation.

Third, Hill-Burton funds were available only to public and private nonprofit hospitals. This produced a marked decline in the number of for-profit hospitals in operation between 1950 and 1965 (Arnold, 1991), demonstrating the power of capital commitment on the direction of growth of the industry.

The Hill-Burton Act of 1946 thus marked the beginning of a long-term pattern in which public funding – with minimal state decision-making – fed the expansion of the hospital sector. It also allows us to glimpse both the power and the limits of government investment in determining the trajectories of the industry as a whole.

On the one hand despite its heavy involvement in the funding of hospital expansion, the government did not establish effective regulatory devices to oversee health care expenditures. At the same time, Hill-Burton illustrates the very important constraints that public capital flows have applied to the health industry. The most overt of these is the failure of the for-profit sector during this period, resulting from its inability to access Hill-Burton funds.

We see, here, the pattern that has been replicated during both the creation of the post-war health care system and in its devolution into crisis: government and private funding streams have fitted themselves into the existing patterns of finance and policy formation; each one transforming the previous pattern. This new reality has then become the template for future action by both government and private actors.

MEDICARE AND MEDICAID

Medicare and Medicaid follow this same sort of dialectical development. The two systems, which pay for direct health care services to the poor (Medicaid) and the elderly (Medicare) were modeled directly, at AMA insistence, on the already operating fee-for-service, full reimbursement, system developed and maintained by Blue Cross. Unlike the private medical insurance that inspired them, however, Medicare and Medicaid legislation submitted to pressure from the American Hospital Association, and included capital expenditures as reimbursable expenses and built in extremely generous depreciation schedules for those purchases (Light, 1996). Since fee-for-service reimbursement was a cornerstone of the capital formation process in health care until this time, state entry into health purchasing became de facto support for the capital formation policies that had already become institutionalized. But the magnitude of government contributions and the inevitably expanding nature of entitlement programs, coupled with the new incentives for capital expenditures meant that they had a transformative impact on the health care industry.

The state thus made explicit what had been only implicit in the earlier system; and, in the process, created what has been called a blank check for private hospital expansion. The availability of capital attracted a wide assortment of large and small companies to the field, as health care supply and provision became a more and more important source of corporate profits. The vast expansion of government financial participation did not, however, increase government involvement in investment decision-making. Once again, the government fit itself into the niche already allocated to insurance carriers; and the role of the state in policy formation was limited to terms of payment, thus enabling the private sector to continue operating autonomously (Arnold, 1991).

But even without altering the pattern of decision-making, the sheer magnitude of the government funding stream associated with Medicare and Medicaid wrought major consequences in the nature of health industry capital formation, notably a dramatic shift in sources of capital (Cohodes & Kinkead, 1984). By substantially increasing the already significant proportion of hospital costs paid by third parties, and by making capital costs explicitly reimbursable, these programs made hospital revenues secure enough (in the eyes of financial markets) to be reliable risks for debt financing[3] (National Journal, Feb. 5, 1977). These changes in financing became the structural backdrop for the sustained period of technological innovation that has characterized the health care industry since the early 1970s. By the late 1970s, the revolution in medical technology resulting from a capital source able to implement any and all technological innovations began to have a recursive effect on the financing arrangements of medical facilities, and on the competitive environment which drove their actions. Debt financing, which only a decade or so earlier was a newfound luxury that allowed for new development and broadened research horizons, became a necessary part of the yearly routine. The vast and continuing innovation in capital intensive technology soon became a prerequisite for hospital operation. The life cycles of new equipment became shorter and shorter as depreciation schedules encouraged the state-of-the-art equipment that became more and more central to the competitive strategy of providers. Thus, the pace of change and the cost of new equipment created an inflationary spiral in which escalating costs of new equipment made debt financing a necessary part of competitive upgrading.

These processes completely transformed the profile of capital flows into the health care sector. By 1983, about 70% of hospital construction capital came from private sector loans (Cohodes & Kinkead, 1984), compared to less than 10% only 35 years earlier. Moreover, unlike the federal government, private lenders were not content to be passive investors; instead of modeling their involvement after existing patterns, the financial community adopted the policies they practiced with other business loans. The insurance industry, in particular, became the major source of construction capital, and therefore a major decision maker requiring competitive business practices that determined the developmental and policy trajectories of the health delivery system.

Here, again, we see the replication of the pattern established by the Hill-Burton Act. The federal government, through its creation of Medicare and Medicaid, simultaneously maintained and transformed the health care system. In this instance, by vastly expanding the system of reimbursements developed by nongovernmental suppliers of health insurance, it transformed the financing system by providing access to vast amounts of internally generated capital that had never before been available to health care companies. By introducing reimbursement for capital expenditures and encouraging rapid turnover via depreciation, it set in motion a series of transformative processes. The spiral of innovation that has revolutionized health care technology could not have occurred without these developments, and this spiral combined with heightened competition among hospitals to create ever shortened life cycles for new equipment, spiraling costs, and a growing dependence on debt financing as an everyday part of the hospital business. And, finally, this dependence brought the industry under the thrall of hegemonic financial institutions, which treated health care as a profitable target for investment. Decisions about the location of investment capital had profound effects on the environment in which hospitals and other health care organizations functioned.

Ultimately, the "hands off" policy of the federal government with regard to industry decision-making, which preserved the autonomy of health care and left major strategic choices in the hands of health care professionals, was undermined. The dynamics triggered by Medicare and Medicaid led large financial institutions, insurance companies, especially, to become key players in strategic decision-making, and this ultimately undermined the autonomy that federal policy had been designed to protect.

THE RISE OF THE PRIVATE MULTI-HOSPITAL SYSTEM

While most histories of private sector multi-hospital systems root their development in the reimbursement provisions and reporting requirements of federal funding programs discussed above, the role of the private capital market is in many ways more important. Medicaid and Medicare contributed to the rapid growth of proprietary hospital chains in the same way that it encouraged debt financing. By guaranteeing payment for services for an additional sector of the population, it made health care an attractive outlet for private investment. It also provided for generous returns on equity investment (Light, 1986) and, in doing so, created the conditions for profitable operation by for-profit facilities. Thus, the 1970s had largely reversed the dampening effects of the Hill-Burton Act on private hospital development, and the private sector began to thrive.

As with other federal actions, the resuscitation of the for-profit sector set in motion dynamics that would transform important aspects of the system. In this case, the transformative force was the rise of privately owned multi-hospital systems. While in 1961 there were only five multi-hospital systems of any type in the United States, by 1985 the figure had grown to include 53 investor-owned chains, controlling more than 680 hospitals and over 100,000 beds (Starr, 1982; Light, 1986). Moreover the growth rate of the largest companies was even more dramatic than the rest of the sector. In 1975 the four largest companies together grossed less than $950 million; by 1981, each generated over $1 billion, representing an annual (compounded) rate of growth of 37.4% (Cohodes & Kinkead, 1984). And this differential development led to the sort of concentration that is commonplace in other industrial sectors: by 1985, one company, Hospital Corporation of America (HCA), held one-half of the beds owned by multi-hospital firms, establishing itself as the General Motors of health care (Light, 1986). More recently, its revenue base has grown to almost 20 billion dollars (*Fortune Magazine*, 1999).

The rise of the multi-hospital systems marked the entree into health care of the nexus of forces that characterize private capitalism. Certainly the array of government actions we have reviewed played a key role in laying the foundation for these national firms, but their rise to dominance reflected the routine market-driven financial advantages of large firms in most industrial and commercial sectors. In particular, these firms could systematically exploit the high inflation levels of the 1970s to service vast amounts of long-term debt in cheaper dollars, and this rewarded strategies of acquisition and expansion. And, once they became sufficiently large to enter the established capital markets, the very high profits of the industry made them the darlings of Wall Street, and therefore gave them access to even greater resources (Salmon, 1985). In short, the broader context of the American economy worked to give large multi-hospital companies a decided advantage in the health care industry.

This emergence of large, dominant, firms that could compete with established companies in other sectors for the favor of Wall Street finally consolidated the role of financial firms in the ongoing planning of the health care system. Since corporate policy in these firms was driven by bottom line considerations, their primary goal was to maintain their stock and bond prices and please their lenders. Life insurance companies, in particular, played an important role in this process as major sources of hospital construction capital. In the late 1970s, for example, Hospital Corporation of America had borrowed more from the insurance industry than from any other source: $173.5 million to finance 39 projects, as compared to $36.4 million outstanding in industrial revenue bonds, the second largest type of creditor.

As early as the mid-1970s, the consequences of this funding for health care delivery were becoming apparent. William Duffy, Assistant Manager of Equitable Life, nicely illustrated these consequences when he commented that "lenders . . . do not want to make an investment in an unprofitable . . . proposition . . . such as hospitals in the inner city" (quoted in The National Journal, Feb. 5, 1977). We will return to this point shortly.

In the midst of the rise of large firms in the hospital industry, government was itself responding to the same realities of inflation and industrial crisis by seeking to limit the escalating costs of medical entitlement programs. The introduction of the diagnostic-related group billing system (DRG), signed into law in 1983 by President Reagan, replaced the traditional fee-for-service reimbursement procedures of Medicare and Medicaid with a system of fixed price reimbursement dependent upon specific diagnoses. As in past cases, the shift in government policy had vast unintended consequences for the shape and dynamics of the industry.

Had this change taken place before the rise of the multi-hospital firms, these companies – and the entire for-profit sector – might never have come to dominate the industry, since fee-for-service medicine, with their generous debt financing reimbursements, had been the engine of privatized finance. But since this sea change in reimbursement policy occurred after they were already large and dominant, the altered federal policy set in motion a search for sectors of the industry that would be profitable without the guaranteed subsidy of cost-plus reimbursement.

In this case, therefore, government action once again had a vast impact on the nature of private sector involvement in health delivery, but the profit nexus now mediated this impact. As Light (1986) points out, the logic of profitability was soon extended to the private nonprofit hospital, since they competed in the same capital markets as the profit making firms, and therefore were required to meet the same return-on-investment criteria as their profit seeking competitors. Therefore, though the for-profit chains accounted for a relatively small portion of total United States hospital capacity, the policies of most hospitals were to a considerable degree driven by the requirements of capital suppliers.

With the profit nexus at the center of industry dynamics, the response to the change in fee determination led to a renewed search for the most profitable use of investment capital. One consequence of this search was the upstream integration that dominated corporate policy during the late 1980s. Another, less visible, change was the mix of patients at private sector hospitals. The proprietary sector made it clear that it felt little obligation to patients covered by Medicaid, since they could gain much higher profits elsewhere; and this logic was extended to nonprofit hospitals by the lenders, who preferred to place their money in ventures with a greater probability of financial viability. By the late 1980s, this trend had extended to beyond the multi-hospital firms and into the nonprofit private sector.

This example of cream skimming exacerbated the larger pattern and was felt most severely in public hospitals, which were still required to take undercovered and uncovered patients. Thus, this process contributed in important ways to the public sector as a dumping ground for the indigent, with public hospitals a larger and larger burden on municipal budgets and increasing targets for budget cutting conservatives who argued that the government should not be subsidizing unproductive members of the population.

We see, then, another round of government actions and industry reactions that reflect the larger pattern we have been tracing, but which also introduce new dynamics that have become central to industry trajectories. On the one hand, government DRGs, like government initiatives in the past, wrought vast changes in industry trajectories, as the for-profit sector (and then inevitably the private non-profits) migrated into profitable areas, leaving unprofitable areas to atrophy or to become political targets in the public system. On the other hand, these actions reflected the arrival of the full dynamics of a capitalist sector: the rise of dominant core firms; the collateral rise of a central decision making network that makes key choices for industry development; the involvement of lenders in this network; and the use of capital flows to extend the decisions taken at this center outward to the rest of the industry.

STATE ATTEMPTS AT CAPITAL CONTROL

As early as the 1970s, health care inflation was recognized as an important problem by the key institutional health care consumers: big business, labor, and the state. Before 1980, however, only the state sought to control health care costs, usually by taking a more activist stance toward the direction and size of state-funded capital flows.

The first important attempt at capital direction was Richard Nixon's Health Maintenance Organization (HMO) strategy, formalized by the Health Maintenance Organization Act of 1973. It encouraged replacing fee-for-service medicine with an organizational form designed to control costs. The vision – one in which the business community had shown substantial interest – was a system of competitive, privately owned, businesses in which physician-employees would practice price conscious medicine reinforced by controls embedded in the new organizational structure (Leyerle, 1994).

Corporate interest in the idea can be readily understood when we consider the often quoted assertion by Paul Ellwood that HMOs "could stimulate a course of change in the health industry that would have some of the classical aspects of the industrial revolution – conversion to larger units of production, technological innovation, division of labor, substitution of capital for labor, vigorous competition, and profitability as the mandatory condition of survival" (Ellwood, 1971: 291).

While the plan included capital subsidies to private industry as seed money for development, it looked toward the private sector for investment capital for development. And, while private industry seemed quite interested – government documents of the time period listed 14 major corporations, twenty entrepreneurial organizations and 15 insurance companies actively exploring HMO investment opportunities – development was hampered by a series of governmental and market factors (Salmon, 1984). Concurrent cutbacks in government health funding made health care less attractive than earlier times as an investment site; other government regulations designed to contain costs inhibited rapid growth; and the recession of 1973–1975 limited the amount of private capital available for new investment. When conditions changed in the early 1980s – including the repeal of important health planning legislation spearheaded by the Reagan administration – private industry again became very interested in HMO investment possibilities.

By the late 1970s the failure of Nixon's HMO strategy to attract private capital led the Carter administration to adopt a very different approach to health care cost containment: an attempted moratorium on all direct and indirect public subsidies for health construction projects. Totaling $4.9 billion dollars in 1977, the proposed moratorium was viewed by many as an effective way of shrinking the health care system (Iglehart, 1977a). The compromise $2.5 billion spending cap that was considered by Congress had the support of a surprising number of corporate interests that were intent on cost control. It was clear, however, that even this compromise would force major changes in the private health care capital market; and since major lenders, commercial banks, pension funds and insurance companies had all come to depend upon these subsidies to guarantee their very large exposure in the industry, they were therefore opposed to the measure. (Iglehart, 1977b). Ultimately, the bill became the victim of a massive lobbying effort by the hospital industry (Starr, 1982), but this conceals the key role played by the financial community. Without the support of the lenders, those firms that sought to support the moratorium were left without a key ally, and this undermined their efforts to overcome the overt opposition of the medical industry.

Ronald Reagan's posture echoed Nixon's encouragement of private sector entry into health delivery, emphasizing the development of HMOs specifically. In 1981, for example, the administration organized a one-day conference designed to encourage private investment in HMOs, inviting 150 representatives of insurance companies, investment houses, large corporations and venture capital firms (Newsweek, Oct. 12, 1981). And, indeed, the 1980s brought rapid growth in private HMO development.

The most significant challenge to business as usual, however, came from tighter controls on public spending in the form of the diagnostic-related group billing system mentioned above. DRGs were designed specifically to limit government funding of health care and, in doing so, the government at last became an active purchaser of health services, required for the first time to look at the details of services offered and to evaluate both their medical and financial viability. (Ruggie, 1992).

Undoubtedly, DRGs had an important effect on the U.S. health delivery system, including reduced utilization, lower health care costs, and standardizing and regulating physician practice. Perhaps the most important result, however, was its effect on the hospital industry noted earlier: changes in reimbursement patterns and hence the flow of capital to hospitals had a direct impact on both hospital expansion and corporate development. But it is important to note that the government was not the sole initiator of these changes; reimbursement patterns of nongovernmental insurance carriers had already begun changing and, private insurance coverage was a more important force in health care finance. Even before DRGs, Medicare, the major purchaser of hospital services, accounted for only about a third of all hospital revenues (Arnold, 1991).

Nevertheless, at first it looked as if DRGs marked a significant shift in state influence over the details of health care development (see for example, Ruggie, 1992). Recent events have demonstrated, however, that Arnold (1991) was correct when she suggested that the most significant change in hospital governance between 1965 and 1985 was the shift to market control. DRGs simply exacerbated that trend; their major impact – as we noted above – was to encourage a change in investment strategy toward vertical integration and toward a more drastic policy of cream skimming in hospital construction and acquisition.

The failed attempts at cost control constitute the denouement of the evolution of health care as a market driven industry. Government action in the past – notably the Hill-Burton Act and the passage of Medicaid and Medicare – had followed the pattern of interventions that sought to preserve existing patterns in the industry, but actually created transformative forces that left the sector drastically altered. The cost savings measures of the 1980s,on the other hand, operated from a different reality. Government action was specifically motivated by the desire to intrude into the dynamics of the industry, and it constituted government action on behalf of itself and on behalf of other sectors that were affected by the industry, the business community most specifically. This generated opposition by the industry itself – a sharp departure from previous initiatives that were welcomed – (at least in retrospect) and strong reservations by the lenders, who were, by the early 1980s heavily invested in the sector's status quo. We therefore witnessed the familiar dynamic of the government as mediator of interests among various economic fractions, rather than as an overpowering force whose actions are the key factors in sector development. In short, health care as an industry had come of age, and policy toward it came to resemble government relations with other large sectors that have a significant impact on economic trajectories, like automobiles, airlines, and petroleum.

THE RISE OF THE HMO

The relation between market forces and state regulation in the United States health care system has become still more clearly defined in the 1990s. While the organizational form of note of the 1980s was the proprietary multi-institutional system, the current era is surely the decade of managed care, with an emphasis on HMOs of various stripes.

The term "managed care" has been used to refer to any and all strategies to contain health care costs (Reuter, 1990). By 1990, 95% of individuals covered by employer based insurance plans were enrolled in some form of managed care including HMOs; traditional fee-for-service programs with added utilization review; and collections of physicians in independent practice who contract with insurers to provide care at discounted rates: Preferred Provider Organizations (PPOs) (Bodenheimer & Grumbach, 1995). HMOs, in particular, have become increasingly popular. By 1990, 39 million Americans belonged to HMOs, up from 9 million in 1980. It is important to note that this growth was predominantly in the for-profit sector: in that year two-thirds of all operating HMOs and every newly formed HMO was a for-profit operation (Bodenheimer & Grumbach, 1995).

Moreover, the HMO sector replicated the same growth pattern as the multi-hospital sector 10 years earlier. In the early 1990s, the 10 largest firms owned about 63% of all U.S. HMOs and accounted for about 70% of the managed care market (Gordon et al., 1994). Since then two of the largest providers, Metropolitan Life and Travelers, have merged (Woolhandler & Himmelstein, 1994). In 1992, there were 48 different mergers or acquisitions among HMOs (Bodenheimer & Grumbach, 1995) and more recently, turned into vertical integration: the trend for the late 1990s is an integrated network of HMOs, hospitals, and contracted physicians owned by a for-profit insurance company or multi-hospital system. By 1999, Aetna had become the nation's largest health care provider and managed care company (Best's Review, 1999).

This structural integration represents the development of American health care as a capitalist sector characterized by a familiar structure and following a routine industry dynamic. Indeed, in the early 1990s, the managed care industry was remarkably profitable by the standards applied to all industries. CIGNA, one of the largest American insurance companies and a leader in vertically integrated health systems, earned $234 million in profits in 1993 with its managed care business its strongest performer; for Travelers, another insurance company, managed care products grossed $3.6 billion in 1992, accounting for about 37% of total revenues (Gordon et al., 1994). It was this profitability that has driven corporate expansion into health delivery. In 1994, the United States Congress rejected the Clinton Administration's health care reform proposal declining to implement so dramatic a change in the organization of the health care system. While Congress debated, however, the penetration of managed care into health care delivery was producing changes that were as dramatic as Clinton proposed.

As managed care became the accepted form for the most profitable firms, it became the industry template enforced by the decisions of the financial community. As a consequence, it has become increasingly difficult for any hospital, nonprofits included, to avoid the managed care model, since that is precisely what creditors were demanding. The AMBAC Indemnity Corporation, for example, one of the country's leading tax-exempt bond insurers, specifically evaluated a hospital's future as a managed care sight in evaluating all bond issues. The evaluation is remarkably detailed: they looked at the proportion of the hospital's billings in managed care, the geographical dispersion of its services, the level or profitability, and the degree to which physicians have been integrated into the system (The Bond Buyer, 8/9/93). In short, the proprietary, vertically, integrated managed care network fast became the institutionalized form of health care delivery for the United States in general. As in all economic sectors, however, managed care too is subject to the dynamics of the market. And as health care purchasers, both public and private, continue to squeeze out cost savings, the late 1990s have brought a crisis in managed care profitability. We now see the beginning of a search for a new institutionalized form and we believe that the role of capital will be decisive in the development of the next generation of health care companies.

CONCLUSION

The intervention of the U.S. Federal government into the dynamics of American health care has been at the center of industry trajectories, but private-sector investment decision making have also had profound effects on the development of the U.S. health delivery system. The public and private sectors have interacted, such that each decision shaped subsequent actions. The result has been a transformation that has been unintended and, perhaps, unwanted. The peculiar "hands off" approach that initially characterized post-war government initiatives did not result in the light touch that policy makers intended. By the time Congress moved to regulate the health care industry it had, instead, become a central force in shaping the 'marketized' industry that entered the 21st century.

In some very broad sense, however, these efforts have been a major success. Beginning with the Hill Burton Act, we can discern an underlying logic to government activity that seems to make sense out of the apparently uncontrolled dynamics that we have just reviewed. The construction funds provided by Hill-Burton were a substitute for the sort of investment capital that private lenders provided for marketized industries. Thus, we understand the Hill Burton Act as a functional substitute for private funding when the industry was not seen as a promising locus of investment, either by Wall Street or by local finance. Medicare and Medicaid had a similar profile: they constituted a substitution of the government for the health insurance companies, who were not motivated by the market to service these groups (at a time when they had sufficient political leverage to mobilize government concern).

These two government initiatives were the central elements that set in motion the marketization of health care and thus, the entrance of corporate capital into health care delivery. In these cases it was the influx of capital – public capital, more specifically – that drove development. As health care evolved into an industry in the American capitalist context, it became a site for large amounts of private investment capital and for the growth and consolidation of large, dominant, firms that could set the standards and strategies of the broader industry. The exposure of large financial firms, particularly insurance companies, guaranteed both continuing access to financial markets, as well as the enforcement of profit maximizing strategies on the fragmented smaller firms.

The force of these dynamics is revealed by the reaction of the industry to the potentially devastating enactment of DRGs. Though this change undermined the foundation of prior growth, it did not drive profit-seeking corporations from the industry. Their involvement and the involvement of the lenders who financed them had become irreversible; and they therefore responded strategically – by attempting to capture and/or create highly profitable sub-markets. Much of the current health care crisis derives from the success of this strategy, as the largest firms captured the most lucrative businesses and left the mundane (and unprofitable) jobs (like supplying health care to the poor and other financially insecure populations) to the smaller private firms and – most dramatically – to the overburdened and underfunded public sector.

This new development also fundamentally altered the ways in which government could intervene in the trajectory of the industry. Whereas a fragmented and dependent health care system in the 1940s was grateful for whatever help it could get – particularly since the government was happy to leave the internal decision making alone – the marketed health care industry is large and well organized, and therefore capable of creating powerful political blocs whenever government policy threatens their interests. This new activism is a two edged sword. It implies the ability of the industry to intervene in government policy formation to prevent the passage of laws or the inclusion of programs that are unpalatable to the core players as in the recent case of the Patients' Bill of Rights. It also implies that the industry – at least its key institutions – can utilize normal Washington politics to mobilize government resources when they are needed. In short, the health care sector has joined the ranks of those industries that utilize access to government as a part of their repertoire of profitability.

The government remains important to industry trajectories as it continues its attempts to control its costs. This is illustrated by the current retraction of Medicare HMOs. It is no longer at the center, however. Like other major marketed industries, the future of health care will be determined by the broad forces of the market – including reactions to government initiatives – mediated by the decisions taken in the boardrooms of the major players, and constrained by the capital flow decisions of both the financial community and the government. This will, in turn, impact on subsequent government decisions. Thus the entrance of corporate capital, conditioned by public policy decisions in general and by public capital contributions to health care more specifically, has solidified into a determining force in both health care delivery and health care politics.

NOTES

1. The authors would like to thank Donald Light and Lawrence Weiss for detailed comments and Carol Fitzpatrick for research assistance.

2. The insurance industry is divided into two segments, life/health on the one hand and property/casualty on the other, with life/health dominating. In 1989, for example, insurance companies sold 13 times as much life and health insurance as property/casualty (Weiss, 1992).

3. While Medicaid and Medicare, with the help of other federal subsidy programs, provided the conditions for debt financing, the vehicle that brought hospitals into capital markets was the state-level tax-exempt bond. Cohodes and Kinkead (1984) suggest that given the extent of federal funding of health care, it is ironic that what has become the major vehicle for capital subsidy came from the level of the 50 states.

REFERENCES

Arnold, P. (1991). The invisible hand in healthcare: The rise of financial markets in the U.S. hospital industry. In: J. Campbell, J. R. Hollingsworth & L. Lindberg (Eds), *Governance of the American Economy.* New York: Cambridge University Press.

Berle, A. Jr., & Means, G. (1932) *The Modern Corporation and Private Property.* New York: Harcourt, Brace and World.

Best's Review – Life Insurance Edition (1999). Aetna to pay $1 billion for Pru's health care unit, 99, no. 9, p. 14. Feb. 2.

Bodenheimer, T. (1990). Should we abolish the private health insurance industry? *International Journal of Health Services, 20,* 199–220.

Bodenheimer, T., & Grumbach, K. (1995). The reconfiguration of U.S. medicine, *The Journal of the American Medical Association, 274,* 85.

Cohodes, D., & Kinkead, B. (1984. *Hospital Capital Formation in the 1980s.* Baltimore: The Johns Hopkins University Press.

Ellwood, P. (1971). Health maintenance strategy. *Medical Care, 9,* 291.

Fortune Magazine (1999). *The Fortune 500, 139*(8), April 16.

Galbraith, J. K. (1967). *The New Industrial State*. New York: New American Library.

Ginsberg, E. (1998). The noncompetitive health-care market, *Challenge*, *41*(4), 31ff.

Glasberg, D. & Schwartz, M. (1983). Ownership and control of corporations. *Annual Review of Sociology*, *9*, 311–332.

Gordon, S., Shindul-Rothschild, J., Woodall, P., & Watzman, N. (1994). The managed care scam, *The Nation*, *258*, 657.

Health Insurance Association of America (1990). *Source Book of Health Insurance Data: 1990*. HIAA.

Heirich, M. (1998). *Rethinking Health Care: Innovation and Change in America*. Boulder, CO: Westview Press.

Himmelstein, D., & Woolhandler, S. (1994). *The National Health Program Book*. Monroe, ME: Common Courage Press.

Iglehart, J. (1977a). The flip side of Carter's cost control plan, *The National Journal*, *9*, 848–852.

Iglehart, J. (1977b). Capital deformation, *The National Journal*, *9*, 1699.

Kopcke, R. (1995). Financial innovation and standards for the capital of life insurance companies, *New England Economic Review*, Jan.–Feb., 29–57.

Leyerle, B. (1994). *The Private Regulation of American Health Care*. Armonk, N.Y.: M. E. Sharpe.

Light, D. (1986). Corporate medicine for profit. *Scientific American*, *255*(6), 38–45.

Light, D. (1997). The restructuring of the American health care system. In: T. J. Litman & L. S. Robbins (Eds), *Health Politics and Policy*. Albany: Delmar.

Mintz, B., & Schwartz, M. (1985). *The Power Structure of American Business*. Chicago: University of Chicago Press.

Moody's Bank and Finance Manual (1998). New York: Moody's Investors Service.

Reuter, C. (1990). Compensation management in practice, *Compensation and Benefits Review*, May/June, 14–17.

Ruggie, M. (1992). The paradox of liberal intervention: Health policy and the American welfare state, *American Journal of Sociology*, *97*, 919–944.

Salmon, J. W. (1984). Organizing medical care for profit. In: J. McKinlay (Ed.), *Issues in the Political Economy of Health Care*. New York: Tavistock.

Salmon, J. W. (1985). Profit and health care: Trends in corporatization and proprietization, *International Journal of Health Services*, *15*, 395.

Skocpol, T. *Protecting Soldiers and Mothers*. Cambridge, MA.: Harvard University Press.

Starr, P. (1982). *The Social Transformation of American Medicine*. New York: Basic Books.

The National Journal (1977). *How hospitals pay their way*, *9*, 207.

Weiss, L. (1992). *No Benefit: Crisis in America's Health Insurance Industry*. Boulder, CO.: Westview.

Woolhandler, S., & Himmelstein, D. (1994). Giant H.M.O. 'A' or Giant H. M.O. 'B'? *The Nation*, Sept. 19, 265–267.

THE INFLUENCE OF MID-LIFE ADULT STATUS AND FUNCTIONAL HEALTH STATUS ON HEALTH LIFESTYLES

Neale R. Chumbler, Ashley Foster, James W. Grimm and Philip Williams

ABSTRACT

The objective of this chapter was to advance the medical sociology theoretical literature on health lifestyle behaviors. This study investigated the influence of mid-life adult statuses (gender, marital status, parenthood, presence of children in the home, and employment status) and functional health status (general physical health, general mental health, vitality, and social functioning) on selected health-related behaviors. These behaviors included preventive medicine behaviors (routine physical exams, cholesterol checks, and blood pressure checks), risk-taking behaviors (driving above the speed limit, smoking, and consuming alcohol), health promoting behaviors (exercising, sleeping well, and relaxing), and medication usage (using prescribed medicines, over-the-counter medicines, and herbal supplements). The data were gathered through the Warren County Health Survey, a telephone survey of a randomly selected sample of residents from a county in Kentucky. Logistic regression was used to analyze the data. Results indicated that having children at home, being married, being fifty years old or older, and possessing better mental health and social functioning statuses were associated with a greater likelihood of practicing better health-related behaviors. In contrast, being both male and aged 49 and younger were correlated with a greater likelihood of

Health, Illness, and Use of Care: The Impact of Social Factors, Volume 18, pages 249–267.

ISBN: 0-7623-0740-4

engaging in poorer health-related behaviors. Based on these results, the authors offered several testable, theoretical propositions for future research to test the relationships between mid-life adult health statuses, functional health statuses, and health-related behaviors in other community-based samples.

INTRODUCTION

The negative impacts of disadvantaged socioeconomic status in relation to health, morbidity, and mortality are well known. Nevertheless, socioeconomic differences leave much unexplained variation in adults' health-related behaviors and attitudes (Cockerham, Rutten & Abel, 1997; Ross & Bird, 1994). Research in the United States and in Europe has found considerable variation in diet, exercise, drinking, and smoking that crosses class lines (Cockerham, Kunz & Lueschen, 1988). Suggestive clues for potential explanations of the health-related actions and ideas that transcend class have been raised by research showing differences in health-related patterns of attitudes and behaviors by gender, employment status, marriage and divorce patterns, contexts and duration of employment, and subjective health assessments (Mechanic, 1999; Ross & Bird, 1994). However, the interrelationships among these characteristics and different health-related behaviors are not fully understood (Liang, Shediac-Rizkallah, Celentano & Rohde, 1999).

While the importance of health lifestyles and health self assessments is recognized as being significant in relation to morbidity and mortality, the task of determining such effects is generally viewed as being difficult (Cockerham et al., 1997; Mechanic, 1999). On the one hand, the components of health-lifestyles are complex and may be somewhat independent of health self-assessments (Mechanic, 1999). On the other hand, the social factors that have been shown to have the potential to influence health-related behaviors are extensive. A reconsideration of the current conceptualization of health lifestyles, health self-assessments, and the social determinants having the potential to influence adults' health behaviors is necessary in order to better understand the psychosocial contexts of contemporary health among adults (Cockerham et al., 1997; Mechanic, 1999). In an effort to advance medical sociology literature, the present study reassesses the conceptualization of the interrelationships of mid-life adult statuses and functional health status on health lifestyles. It further analyzes the health lifestyle behaviors of adults apart from the confounding effects of socioeconomic status, access to health insurance coverage, and other effects of economic hardship on health status and health-related behavior (Link & Phelan, 1995; Marmot, Bobak & Smith, 1995).

The present study should be interpreted as an exploratory study, where evidence can be garnered about patterns in adult health lifestyles that might be examined in larger, more representative samples. Following the data analysis section, testable propositions are hypothesized.

For three principal reasons, we employ inductive reasoning rather than deductive reasoning when theorizing about the effects of the mid-life adult status set on health lifestyles. First, the paucity of prior research on the composite effects of adult statuses makes predictions tentative. Second, from the very few previous studies, rather inconclusive findings have emerged. For instance, while empirical evidence suggests that the longer term effects of statuses, such as full time employment may improve health status (Pavalko & Smith, 1999), role-related research suggests that the adult status set may exacerbate health status. When an individual has too many roles (i.e. role overload) and/or is captivated in a current role by being an unwilling incumbent of roles attached to statuses, then it worsens health status (Aseltine & Kessler, 1993; Simon, 1992). Third, in part because of the interrelatedness of statuses, such as between employment and gender, the same statuses may have varying effects, depending on how they are experienced and interpreted (Mechanic, 1999).

CONCEPTUAL DEVELOPMENT

Adult Statuses

While some combinations of adult statuses and temporal patterns of activity related to them have been studied in relation to health, such as the contexts and duration of employment by gender in relation to objective health-related measures (Ross & Bird, 1994; Pavalko & Smith, 1999), the implications of mid-life adult statuses as a whole for health have not yet been studied (Mechanic, 1999). No previous research has addressed the potential influence of adults' *mid-life adult status set* on patterns of health-related behaviors. The mid-life adult status set includes the following social variables: gender, marital status, parenthood, presence of children in the home, employment, and civic involvement (Hess, Markson & Stein, 1996, p. 119). This dearth of information is disconcerting, since the status set expands from childhood to mid-life and shrinks with disruptions of social interrelationships that accompany aging (e.g. retirement; widowhood) (Hess et al., 1996, p. 76). Consequently, the mid-life adult status set could exert considerable influence on health-related behaviors.

Functional Health Status

Previous research has shown the independent role of health perceptions in relation to objective health indicators as well as to health behaviors. A long tradition of research shows that subjective health appraisals, psychological distress, and judgments of self-functioning in association to health are quite important in relation to health-related behaviors (Cockerham, Sharp & Wilcox, 1983; Mechanic & Angel, 1987). While the perceived seriousness of symptoms has been found to be strongly related to health behaviors, such as consulting physicians (Berkanovic, Telesky & Reeder, 1981; Tanner, Cockerham & Spaeth, 1983), the roles of health self-assessment and psychological distress in relation to adults' health lifestyles have yet to be conceptualized or adequately studied (Cockerham et al., 1997). Research suggests that health beliefs and especially vulnerability may play a general role in health life-styles and preventive behavior with regards to dietary control among obese children (Becker, Maiman, Kirscht, Haefner & Drachman, 1977), self-care behavior (Haug, Akiyama, Tryban, Sonoda & Wykle, 1989), and management of hypertension (Brown & Segal, 1996).

Health Lifestyles

Health lifestyles are collective patterns of health-related behaviors people chose from the options available to them, considering their life chances as impacted by socioeconomic status, age, and other factors (Cockerham, 1998, p. 85). In addition to annual physical exams and taking prescribed medications, health lifestyles include decisions people make concerning risk-taking behaviors (e.g. driving above the speed limit), health promoting behaviors (e.g. exercising), and medication usage (using over the counter [OTC] medications) (Cockerham, 1998, p. 85). These behaviors have improved health status, lengthened life expectancy, and reduced the risk of mortality from heart disease and other chronic diseases (Lee, Hsieh & Paffenbarger, 1995; Shepard, Cobbe, Ford et al., 1995).

Following commonly used components of lifestyle behaviors in previous research, we included several types of behaviors in conceptualizing adult health lifestyles (Donelan, Blendon, Hill et al., 1996; Liang et al., 1999). Health screening behaviors included routine physical exams, cholesterol checks, and blood pressure checks. Risk-taking behaviors included driving above the speed limit, smoking, and consuming alcohol. Health promoting behaviors included sleeping well, exercising, and relaxing. Medication usage included using the following: (1) prescribed medications, (2) OTC medications, and (3) herbal/vitamin supplements.

METHODS

Sample

The data are from the 1998 Warren County Health Survey (WCHS), a telephone survey of adults (age 18 or older) living in Warren County, Kentucky, conducted by the Western Kentucky University Social Research Laboratory, that focuses on health-related concerns. Warren County is defined as a non-metropolitan statistical area county containing approximately 80,000 people. The county represents individuals who tend to be relatively well educated, employed, and moderately wealthy.

The sampling design is a probability survey employing a random digit dialing technique. The researchers obtained a county-based list of valid telephone area codes and prefixes followed by four random digits. This type of sampling frame represented all telephone households (it did not distinguish between directory-listed and unlisted numbers) and guaranteed that every residential telephone line in the county had an equal probability of being selected. This method yields an almost totally unbiased sample of households with telephones (Chumbler, Hartmann, and Beck in press; Mainous, Bertolino & Harrell, 1992).

Households were contacted from 6:30–8:30 p.m. on Monday through Thursday. At least four attempts were made to contact every number within the sample. If respondents were reached at an inopportune time, callback appointments were made. The response rate (percent of completed interviews out of the total number of contacted eligible households) was 48.0% (refusal rate = 52%), with data provided by 182 individuals. The refusal rate is comparable to other national surveys. The Council for Marketing and Opinion Research (1998) found, on average, a refusal rate of 58% for surveys conducted in 1997 with the general public using a random-digit dialing methodology sample and surveys lasting 10–20 minutes in length.

Table 1 presents a comparison of selected characteristics of the WCHS sample to the 1990 U.S. Census population for Warren County, KY. As can be seen, the respondents to the WCHS were significantly better educated and more likely to be employed. Moreover, respondents to the WCHS were somewhat more likely to be female. With regards to gender, ethnicity, age, and marital status, respondents to the WCHS were representative to the 1990 U.S. Census population characteristics for Warren County, KY.

Measures

Dependent variables

We considered four categories of dependent variables, which pertained to health lifestyle behaviors: risk-taking behaviors, health promoting behaviors, preventive medicine behaviors, and medication behaviors. *Risk-taking behaviors* included driving above the speed limit, smoking, and consuming alcohol. *Health promoting behaviors* included exercising, sleeping well, and relaxing. *Preventive medicine behaviors* included routine physical exam, cholesterol evaluation, and blood pressure evaluation. *Medication behaviors* included OTC medications, prescription medications, and herbal medications. To investigate characteristics of adults who demonstrated these health lifestyle behaviors we dichotomized the respondents into those who did and those who did not display these behaviors everyday or most days *one month* prior to the interview.

Table 1. Comparison of Selected Characteristics of the 1998 Warren County, KY Health Care Survey to the 1990 U.S. Census Population for Warren County, KY

Variable	Warren County, KY Health Care Survey %	Population Characteristics for Warren County %
Gender		
Male	41.9	47.9
Female	58.1	52.1
Ethnicity		
White	87.4	90.3
Non-white	12.6	9.7
Age		
18–24 years	14.9	20.6
25–44 years	45.3	40.7
45 + years	39.8	38.7
Education attainment[a]		
High school graduate or higher	95.5	70.9
Marital status		
Married	57.7	55.8
Labor force status		
In labor force[b]	73.6	66.3

[a] for persons 25 years and over.

[b] includes full-time, part-time and self-employed.

Independent variables

We considered two categories of independent variables: adult status and functional health status. *Adult status* variables were intended to reflect the maximum composite of adult commitments and responsibilities in mid-life (Hess et al., 1996, p. 119). These variables included gender (1 = *male*), marital status (1 = *married*), having children 18 years of age and younger living at home (1 = *yes*), and employment status (1 = *employed full-time*). Age was a continuous variable. However, we created two dummy variables, with an 18–29, 30–49, and ≥50 age grouping categories, with the 30 to 49 age group serving as the referent category. These coding schemes were intended to measure the key effects of status responsibilities in adult mid-life (Hess et al., 1996, pp. 156–157). The race/ethnicity of respondents was included in order to control for the potential effects of it on health lifestyle behaviors.

To assess the *functional health status* of respondents, we employed four subscales. One of the subscales, social functioning, contained the exact number and wording of the items from the Medical Outcomes Study 36-item short form (SF-36) health survey (Ware & Sherbourne, 1992). The other three subscales – physical functioning, general mental health, and vitality – contained a limited number of items from the original SF-36. Due to space and time limitations, as well as through pilot testing of the items, we employed the abbreviated versions of these measures. Our purpose was not to replicate the actual wording of the SF-36 items, but use those items that were most suitable to our sample. Scores for each subscale were the sum of the scores for the questions within each subscale. Higher scores meant better health state (Lorig, Stewart, Ritter, Gonzalez, Laurent & Lynch, 1996).

Social functioning was measured with a two-item measure that examined the extent to which respondents' health or emotional problems interfered with social activities. Scores ranged from 0 to 8 ($M = 7.27$, $SD = 1.63$). Cronbach's alpha reliability with this sample was 0.84. Physical functioning subscale contained nine items that examined the extent to which a respondents' health limited activities such as running or heavy lifting, moving a table or pushing a vacuum, lifting or carrying groceries, climbing stairs, bending, kneeling or stooping, walking one block, walking several blocks, bathing or dressing oneself, and performing normal work outside the home. Scores ranged from 0 to 36 ($M = 30.84$, $SD = 8.47$) and had an internal consistency reliability of 0.93.

The three-item general mental health subscale focused on the extent to which respondents experienced feeling nervous, calm and peaceful, and happy in the previous month (range 2–12; $M = 8.08$, $SD = 2.34$). The Cronbach's alpha reliability was 0.70. Also, the three-item vitality subscale focused on the extent to which respondents' felt full of energy, exhausted, and tired in the month prior to the interview (range 0–12; $M = 6.12$, $SD = 2.59$). Cronbach's alpha reliability was 0.71.

PLAN OF ANALYSIS

Logistic regression analysis (backward elimination) was used to examine the influence of mid-life adult statuses (gender, marital status, presence of children in the home, and employment status) and functional health status (general physical health, general mental health, vitality, and social functioning) on the likelihood of engaging in particular health-related behaviors. Factor analysis indicated that the three *medication behaviors* should be combined into a composite index. Therefore, respondents who used one or more of the three medication behaviors were coded as "1" and those that did not use any of the three were coded as "0". The remaining health-related behaviors were treated separately, with individual logistic regression models employed for each behavior. Each of the models presented in Tables 2–5 were the final models from the backward elimination procedures.

Since the critical purpose of the study is to advance a theoretical perspective, as theory in the area of the current study was not well developed, and because the number of cases was small relative to the number of explanatory variables suggested in the literature, we employed the backward logistic regression procedure (Menard, 1995; Wofford, Elliott & Menard, 1994). Backward elimination begins by inserting all the independent variables under consideration in the model. Then, it deletes one at a time from the model until attaining a position where the remaining variables all make significant partial contributions to predicting health related behaviors (Agresti & Finlay, 1997, p. 529). Backward elimination is the preferred procedure because there is less risk of failing to find a relationship when one exists (Menard, 1995). Following previous research (Menard, 1995; Wofford et al., 1994), the usual 0.05 criterion for statistical significance was relaxed to 0.10.

The odds ratios (ORs) were reported (cf. Tables 2–5). ORs indicated the net change in the likelihood of a respondent engaging in a particular health lifestyle given a one-unit increase in the exogenous variable. A unit increase in the continuous variables (for instance the SF-36 subscales) was one standard deviation. For a binary independent variable, such as gender, the OR indicated the difference in odds for females versus males. Missing values were excluded from the analysis, leaving a total sample size for the multivariate analysis to be 164.

RESULTS

The majority of the study sample was female, married, white, and currently employed full-time. The average age of respondents was 41.2 (range: 18–81). Approximately one-half (49%) of the respondents were age 30–49. Respondents' educational attainment varied considerably, with 46% reporting more than a high school diploma (or equivalent). Individuals were queried whether or not they engaged in certain health lifestyle practices. The sample reported the following behaviors every day or most days in the month prior to the interview: (1) drove above the speed limit (62%); (2) smoked (37%); (3) consumed alcohol (30%); (4) exercised (73%); (5) slept well (85%); (6) relaxed (82%); (7) had a routine physical exam (69%); (8) had cholesterol tested (62%); (9) had blood pressure checked (85%); (10) took medication behaviors (83%).

Multivariate Analysis

Risk-taking Behaviors

Table 2 presents the ORs obtained when each of the risk-taking behaviors (driving above the speed limit, smoking, and consuming alcohol) were regressed on the independent variables. Gender and age were significantly associated with driving above the speed limit. Men were 2.9 times more likely than women to drive above the speed limit. Relative to the 30 to 49 age group, persons in the 18 to 29 age group were 3.1 times more likely, while persons greater than age 50 were 55% less likely to drive above the speed limit. Age and health status were associated with smoking. Persons greater than age 50 were 71% less likely to smoke than those in the 30 to 49 age group. Each one-unit increment on both the general mental health and social functioning subscale scores corresponded with a decrease in the likelihood of smoking by 12% and 19%, respectively. The likelihood of alcohol consumption was significantly associated with gender, age, and health status. Men were 3.2 times more likely than women to consume alcohol. Relative to the 30 to 49 age group, respondents in the greater than 50 age group were 82% less likely to consume alcohol. Each one-unit increment on the general physical health subscale was associated with a 6% increase in the odds of consuming alcohol.

Health Promoting Behaviors

Table 3 presents the ORs of each of the health-promoting behaviors regressed on the independent variables. As regards to exercise behaviors, the likelihood of exercising was significantly associated with employment status, age, and the

Table 2. Logistic Regression Analyses of Risk-Taking Behaviors (Stepwise Regression: Backward Elimination) ($n = 164$)

	Driving Above Speed Limit			Smoking			Alcohol Consumption		
	Parameter Estimate	Odds Ratio	*p*	Parameter Estimate	Odds Ratio	*p*	Parameter Estimate	Odds Ratio	*p*
Male	1.064	2.899	0.005				1.177	3.243	0.002
Age 18–29[a]	1.134	3.108	0.017						
Age 50+[a]	–0.808	0.446	0.067	–1.233	0.292	0.015	–1.744	0.175	0.008
General mental health[b]				–0.131	0.877	0.083			
Social functioning[b]				–0.209	0.811	0.071			
General physical health[b]							0.060	1.062	0.091

[a] referent category: age 30–49.
[b] high scores indicate better health.

Table 3. Logistic Regression Analyses of Health-Promoting Behavior (Stepwise Regression; Backward Elimination) ($n = 164$)

	Exercise			Sleep Well			Relax		
	Parameter Estimate	Odds Ratio	*p*	Parameter Estimate	Odds Ratio	*p*	Parameter Estimate	Odds Ratio	*p*
Male				–1.134	0.322	0.068			
Employed full-time	–0.775	0.461	0.074						
Age 18–29[a]	1.146	3.145	0.019						
Age 50+[a]							1.700	5.473	0.044
Vitality subscale[b]	0.172	1.188	0.014				0.309	1.363	0.004
General physical health[b]				0.079	1.082	0.008			
General mental health[b]				0.575	1.777	<0.001	0.200	1.221	0.059
Married							–0.848	0.428	0.097
Children <18 in home							0.877	2.403	0.078
Social functioning							0.437	1.548	0.041

[a] referent category: age 30–49.
[b] higher scores indicate better health.

vitality subscale score. Respondents who were employed full-time were 54% less likely to have exercised than those not employed full-time. Each one-unit increment on the vitality subscale was associated with a 19% increase in the odds of exercising. Gender and two health status measures were correlated with the likelihood of sleeping well. Men were 68% less likely to have had proper sleep than women. Each one-unit increment in both the general physical health and general mental health subscales was associated with an increase in the likelihood of sleeping well by 8% and 77%, respectively.

Six variables were significantly correlated with respondents' likelihood of relaxing. Persons greater than age 50 were 5.5 times more likely to take time to relax than those in the 30 to 49 age group. Likewise respondents with children under 18 years of age living at home were 2.4 times more likely to take time to relax as those without children less than age 18 at home. Three health status subscales were significantly associated with a greater likelihood of relaxing. Each one-unit increment in the vitality subscale score, the general mental health subscale score, and the social functioning subscale score was associated with an increase in the likelihood of relaxing by 36%, 21%, and 54%, respectively. Finally, persons who were married were 57% less likely to take time to relax than their non-married counterparts.

Preventive Medicine Behaviors

Table 4 presents the ORs of each of the preventive medicine behaviors (routine physical exam, cholesterol check-up, and blood pressure check-up). The likelihood of having a routine check-up was associated with gender, age, and a health status subscale. Men were 50% less likely to have had a routine physical exam than women. Comparative to the 30 to 49 age group, persons in the greater than 50 age group were 2.46 times more likely to have a routine physical exam. Each one-unit increment in the vitality subscale score was associated with an 11% decrease in the odds of having a routine physical exam. Three adult status variables (gender, martial status, and age) were correlated with the likelihood of respondents having their cholesterol evaluated. Men were 44% less likely to have their cholesterol tested than women. Persons who were married were 2.1 times more likely to have their cholesterol checked than those not married. Relative to the 30 to 49 age group, persons in the greater than 50 age group were 8.4 times more likely to have had their cholesterol tested. Marital status and one functional health status measure were associated with blood pressure evaluations. Individuals who were married were 2.7 times more likely to have had their blood pressure tested than their non-married counterparts. Each one-unit increment in the social functioning subscale score

Table 4. Logistic Regression Analyses of Preventive Medicine Behaviors (Stepwise Regression; Backward Elimination) ($n = 164$)

	Routine Physical Exam			Cholesterol Evaluation			Blood Pressure Evaluation		
	Parameter Estimate	Odds Ratio	*p*	Parameter Estimate	Odds Ratio	*p*	Parameter Estimate	Odds Ratio	*p*
Male	–0.599	0.497	0.044	*–0.586*	0.557	.093			
Married				0.747	2.111	0.032	1.000	2.715	.027
Age 50+[a]	0.399	2.456	0.072	*2.133*	8.444	0.001			
Vitality[b]	–0.118	0.889	0.092						
Social functioning[b]							–0.557	0.573	0.080

[a] referent category: age 30–49.
[b] higher scores indicate better health.

Table 5. Logistic Regression Analyses of Medication Behaviors (Stepwise Regression; Backward Elimination) ($n = 164$)

	Medication Behaviors		
	Parameter Estimate	Odds Ratio	*p*
Male	–1.120	0.326	0.010
Age 50+[a]	1.467	4.338	0.057

[a] referent category: age 30–49.

was associated with a 43% decrease in the odds of having a blood pressure evaluation.

Medication Behaviors

Table 5 presents the logistic regression results obtained when the likelihood of medication behaviors (the combination of OTC medications, prescription medications, and herbal medications) was regressed on the independent variables. Gender and age were associated with use of medications. Men were 67% less likely to take medications than women. Relative to the 30 to 49 age group, persons in the 50+ age group were 1.47 times more likely to use medications.

DISCUSSION

This study presented some new information to move forward the medical sociology theoretical literature. The present study sought to identify the extent to which mid-life adult status characteristics and functional health status characteristics influence health related behaviors. We made five major conclusions from the data.

One, we found widespread gender effects on health lifestyles. In the sample of adults, men were significantly more likely to have engaged in risk-taking behaviors (i.e. driving above the speed limit and consuming alcohol) than women. Conversely, women were significantly more likely to have reported preventive medicine behaviors (i.e. obtaining routine physical exams and having cholesterol evaluated), health promoting behaviors (i.e. getting appropriate amounts of sleep), and medication behaviors than men. These results suggest that men may be much less likely to accept responsibility for their own health (cf. Cockerham, 1998, pp. 119–120) than women. Our findings suggest that the increased emphasis given to preventive health behaviors by the mass media, providers, insurance companies and other sources of influence may not be reaching middle-class men. The finding that men in our sample were less likely to have had a routine physical exam or to have taken prescribed medications suggests that men may have received fewer preventive medicine behavioral messages from providers than did women. It is also possible that factors related to gender roles explain both the fewer contacts men have with providers and their failure to engage in important preventive behaviors. Given the sociodemographic profile of the study sample, it seems unlikely that the difference in health lifestyles that we have obtained can be linked to the socioeconomic inequality issues that have surfaced in studies of national samples (cf. Ross & Bird, 1994). The findings propose the following hypothesis which can be tested in future research: *Gender-related factors make men less likely to accept responsibility for their own health, avoid high-risk behaviors, and seek advice from providers about health-related issues than women.*

Two, with the exception of full-time employees being less likely to report that they regularly exercised, our results showed that higher educational attainment and full-time employment were unrelated to respondents' health lifestyles. These findings are inconsistent with past research from national samples that found greater educational achievement and employment status to be associated with greater health status (Pavalko & Smith, 1999; Ross & Bird, 1994). While only suggestive, our results show that when socioeconomic circumstances are more or less equal, and that when people are not suffering economic hardship, differences in health lifestyle may be much less related to differences in either educational attainment or employment. Given the substantial effects of being age 50 years and older (see discussion below), our study suggests that among adults whose socioeconomic circumstances are relatively comfortable, decisions to live more healthy lives are related more to aging and to other social psychological markers (Mechanic, 1999). Thus, our study suggests the following hypothesis that can be used in future research. *Among adults, the effects of education and employment are less important in regard to health lifestyles than are the effects of aging and self-assessed health measures.*

Three, previous research suggested that the positive effects of marriage and children in the home on health lifestyles may be related to the socioeconomic viability of households (Ross & Bird, 1994). We found some substantial, though mixed effects, of family ties on adults' health lifestyles. Married respondents were much more likely to report cholesterol and blood pressure evaluations than unmarried members. On the other hand, married respondents were less likely to have taken time to relax than their unmarried counterparts. Moreover, marital status was unrelated to risk-taking behaviors or to two of the health-promoting behaviors (exercising and sleeping well). These mixed results suggest that the effects of marriage on health lifestyle may depend more on the quality of the relationship rather than on marriage itself, as has been suggested by research on the effects of marriage on mental health (Gove, Hughes & Style, 1983). The convincing means by which the presence of children may impact adults' health lifestyles was shown by the respondents' proclivity to take time to relax. Overall, our findings proffer the hypothesis that could be tested in future research: *Positive marital relationships and the presence of children living at home are related to more healthful lifestyles.*

Four, we found considerable and sizeable age effects on adults' health lifestyles. Relative to those age 30 to 49 years, individuals age 50 years or older were more likely to have engaged in preventive medicine behaviors (i.e. routine physical exams and cholesterol evaluations), health promoting behaviors (i.e. having the ability to relax), and medication behaviors. However, individuals age 50 years or older were less likely than those ages 30 to 49 years to have performed risk-taking behaviors (i.e. driving above the speed limit, smoking, and consuming alcohol). Our findings indicate that reaching age 50 years or older may be a social marker by which adults begin to live healthier lives. While our cross-sectional data cannot directly test such a longitudinal speculation, the pattern to our results suggests a very clear difference in terms of health lifestyles between adults who are age 50 years or older, age 30 to 49, and age 18 to 29. The latter age group, as expected, was more likely to engage in exercise and all three risk-taking behaviors (Cockerham, 1998, pp. 88–89). Further, our research suggests that age 50 years or older may be a particularly important health lifestyle threshold in further specifying the generally positive effects of aging in health lifestyles (Backett & Davison, 1995; Cockerham et al., 1988). Such a threshold may also help explain why among mid-life adult women, longer employment is related to better physical and emotional health (Pavalko & Smith, 1999). Overall, our data offers the following hypothesis to be tested in future research: *Among middle class adults becoming 50 years old is a key marker for improved health lifestyles behaviors.*

Five, in terms of functional health status, the present study found several substantial, positive effects of mental health, physical health and social functioning measures on health lifestyle behaviors. The present study employed two measures of mental health status, overall mental health status and vitality. Overall general mental health status was positively associated with health promoting behaviors (i.e. sleeping well and relaxing). Conversely, overall general mental health status was inversely associated with risk-taking behaviors (smoking). This finding suggests that better mental health status was associated with healthier lifestyles. This association may help explain why self-control over health behavior is such an important factor in sustaining participation in preventative behaviors, such as fitness and walking programs (Kotarba & Bentley, 1988).

Vitality was positively related to a greater likelihood of engaging in exercise and taking time to relax. Vitality was associated with a lower likelihood of having had a routine physical exam. With regards to overall physical health status, we found that greater physical health status was associated with a greater likelihood of consuming alcohol and sleeping well. The role of vitality and physical health status in terms of seeing providers and engaging in preventive health behaviors may help explain why the most common response to ailments is self-care and personal responsibility for one's own health (Haug et al., 1989; Kotarba & Bentley, 1988).

The social functioning measure was associated with a greater likelihood of engaging in health promotion behaviors (relaxing). This measure was inversely associated with the likelihood of engaging in risk-taking behaviors (i.e. smoking), and preventive medicine behaviors (i.e. blood pressure evaluations). Consequently, present results suggest that while positively self-evaluated social functioning can be related to healthful behaviors, it may also be related to and sustain the continuance of some health risk behaviors. In the latter sense, our results support Mechanic's idea that self-assessed vitality is very important in relation to risk behaviors (Mechanic, 1999). The findings support the following hypothesis to be tested in future research: *Positive evaluations of subjective mental and physical health, as well as subjective social functioning evaluations, will be positively related to and sustain some health lifestyle behaviors, but may also be related to lesser rates of seeing providers and to the continuance of some health-risk behaviors.*

In sum, the present study has successfully derived important propositions that must be carefully addressed in future studies of health lifestyle behaviors among adults. Such research is important, since it is often assumed that health lifestyle behavior spreads through and beyond the middle class. At the same time, there is evidence that aspects of lifestyle, such as exercise and health screens, may be in decline (Cockerham, 1998, pp. 92–97). Our study suggests that the general direction of future research should expand the focus of study beyond employment and education to include issues such as gender, family life, attaining age 50 years or older, as well as various subjective evaluations of health status. Age, in particular, should receive major attention in any attempt to interpret variation in health lifestyle among adults.

Results of the present study also highlight the need for longitudinal data on relationships between adult statuses and subjective evaluations of health and health-related behavior. Statuses going beyond work and educational attainment may over time influence the following behaviors: seeing physicians, taking personal responsibility for health, avoiding risk-taking behaviors, and continuing preventative behaviors such as exercise. Moreover, subjective views of physical health, mental health, vitality, and social functioning also have the potential to interact with health behaviors to explain why some adults live healthier lives than others do. Such longitudinal studies must deal with complex interactions between the long term social contexts and subjective views of health status as adults reach 50 years of age or older.

REFERENCES

Agresti, A., & Finlay, B. (1997). *Statistical Methods for the Social Sciences* (3rd ed.). Upper Saddle River, NJ: Prentice Hall.

Aseltine, R. H. Jr., & Kessler, R. C. (1993). Marital Disruption and Depression in a Community Sample. *Journal of Health and Social Behavior, 34*, 237–251.

Backett, K. C., & Davison, C. (1995). Life Course and Lifestyle: The Social and Cultural Location of Health Behaviours. *Social Science and Medicine, 40*, 629–638.

Becker, M. H., Maiman, L. A., Kirscht, J. P., Haefner, D. P., & Drachman, R. H. (1977). The Health Belief Model and Prediction of Dietary Compliance: A Field Experiment. *Journal of Health and Social Behavior, 18*, 348–366.

Berkanovic, E., Telesky, C., & Reeder, S. (1981). Structural and Social Psychological Factors in the Decision to Seek Medical Care for Symptoms. *Medical Care, 19*, 693–709.

Brown, C. M., & Segal, R. (1996). Ethnic Differences in Temporal Orientation and Its Implications for Hypertension Management. *Journal of Health and Social Behavior, 37*, 350–361.

Chumbler, N. R., Hartmann, D., & Beck, C. K. (in Press). Appropriateness in Acquiring a Family Caregiver Interview from Older Adults with Suspected Cognitive Impairment. *Aging & Mental Health.*

Cockerham, W. C. (1998). *Medical Sociology* (7th ed.). Upper Saddle River, N.J.: Prentice-Hall.

Cockerham, W. C., Sharp, K., & Wilcox, J. A. (1983). Aging and Perceived Health Status. *Journal of Gerontology, 38*, 349–355.

Cockerham, W. C., Kunz, G., & Lueschen, G. (1988). Social Stratification and Health Lifestyles in Two Systems of Health Care Delivery: A Comparison of America and West Germany. *Journal of Health and Social Behavior, 29*, 113–126.

Cockerham, W. C., Rütten, A., & Abel, T. (1997). Conceptualizing Health Lifestyles: Moving Beyond Weber. *Sociological Quarterly, 38*, 601–622.

Council for Marketing and Opinion Research (1998)). *Industry Watch.* Port Jefferson, NY: CMOR (Reprinted by permission: (1999)). Industry Survey: CMOR's Respondent Cooperation Audit. In: *Sawtooth News: News from Sawtooth Technologies on Computer Interviewing & Analysis, 15*(4).

Donelan, K., Blendon, R. J., Hill, C. A., Hoffman, C., Rowland, D., Frankel, M., & Altman, D. (1996). Whatever Happened to the Health Insurance Crisis in the United States? Voices From a National Survey. *Journal of the American Medical Association, 276*, 1346–1350.

Gove, W. R., Hughes, M., & Briggs Style, C. (1983). Does Marriage Have Positive Effects on the Psychological Well-Being of the Individual? *Journal of Health and Social Behavior, 24*, 122–131.

Haug, M., Akiyama, H., Tryban, G., Sonoda, K., & Wykle, M. (1989). Self-Care: Japan and the U.S. Compared. *Social Science and Medicine, 33*, 1011–1022.

Hess, B. B., Markson, E. W., & Stein, P. J. (1996). *Sociology* (5th ed.). Boston: Allyn and Bacon.

Kotarba, J. A., & Bentley, P. (1988). Workplace Wellness Participation and the Becoming of Self. *Social Science and Medicine, 26*, 551–558.

Lee, I-M., Hsieh, C-C., & Paffenbarger, R. S. Jr. (1995). Exercise Intensity and Longevity in Men. *Journal of the American Medical Association, 213*, 1179–1184.

Liang, W., Shediac-Rizkallah, M., Celentano, D. D., & Rohde, C. (1999). A Population-Based Study of Age and Gender Differences in Patterns of Health-Related Behaviors. *American Journal of Preventive Medicine, 17*, 8–17.

Link, B. G., & Phelan, J. (1995). Social Conditions as Fundamental Causes of Diseases. *Journal of Health and Social Behavior*, Spec. No., 80–94.

Lorig, K., Stewart, A., Ritter, P., Gonzalez, V., Laurent, D., & Lynch, J. (1996. *Outcome Measures for Health Education and other Health Care Interventions*. Thousand Oaks, CA: Sage.

Mainous, A. G., Bertolino, J. G., & Harrell, P. L. (1992). Physician Extenders: Who is Using Them? *Family Medicine, 24*, 201–204.

Marmot, M., Bobak, M., & Smith, G. D. (1995). Explanations for Social Inequalities in Health. In: B. Amick, S. Levine. A. Tarlov & D. Walsh (Eds), *Society and Health* (pp. 172–210). New York: Oxford University Press.

Mechanic, David. (1999). Conceptions of Health. In: Kathy Charmaz and Debora A. Paterniti (Eds), *Health, Illness, and Healing* (pp. 8–17). Los Angeles, CA: Roxbury Publishing Company.

Mechanic, D., & Angel, R. (1987). Some Factors Associated with the Report and Evaluation of Back Pain. *Journal of Health and Social Behavior, 28*, 131–139.

Menard, S. (1995). *Applied Logistic Regression Analysis* (Series/Number 07–106). Thousand Oaks, CA: Sage.

Pavalko, E. K., & Smith, B. (1999). The Rhythm of Work: Health Effects of Women's Work Dynamics. *Social Forces, 77*, 1141–1162.

Ross, C. F., & Bird, C. E. (1994). Sex Stratification and Health Lifestyle: Consequences for Men's and Women's Perceived Health. *Journal of Health and Social Behavior, 35*, 161–178.

Shepard, J., Cobbe, S. M., Ford, I., Isles, C. G., Lorimer, A. R., MacFarlane, P. W., McKillop, J. H., & Packard, C. J. (1995). Prevention of Coronary Heart Disease with Pravastatin in Men with Hypercholestrolemia. *New England Journal of Medicine, 353*, 1301–1307.

Simon, R. W. (1992). Parental Role Strains, Salience of Parental Identity and Gender Differences in Psychological Distress. *Journal of Health and Social Behavior, 33*, 25–35.

Tanner, S. L., Cockerham, W. C., & Spaeth, J. L. (1983). Predicting Physician Utilization. *Medical Care, 21*, 360–369.

Ware, J. E. Jr., & Sherbourne, C. D. (1992). The MOS 36-Item Short-Form Health Survey (SF–36). I. Conceptual Framework and Item Selection. *Medical Care, 30*, 473–483.

Wofford, S., Elliott, D. S., & Menard, S. (1994). Continuities in Marital Violence. *Journal of Family Violence, 9*, 195–225.

UNCERTAINTY: A KEY CHARACTERISTIC OF CHRONIC ILLNESS AND A MAJOR PROBLEM FOR MANAGED CARE

Ariela Royer

ABSTRACT

In the continuing debate among health care providers, policy makers, and medical sociologists, on reforming the delivery of health care, it is necessary to address the social as well as the medical management of services provided to those whose special health care needs have thus far been inadequately met: chronically ill persons. Through analysis of data collected as part of a larger study concerning the experiences of non-hospitalized chronically ill persons, uncertainty, a key characteristic of chronic illness was identified as causing the most problems to ill persons and to those delivering their health care. This uncertainty factor, a non-medical yet illness related problem was mentioned by all respondents much more than their medical difficulties, financial burdens, and their frustrations with health care professionals. This chapter focuses on the impact uncertainty has for chronically ill persons and particularly, the problems it creates for those treated in managed care programs.

INTRODUCTION

According to my data, the most often mentioned problem in living with chronic illness, seem to center around the experience of "uncertainty." Stigma, and

Health, Illness, and Use of Care: The Impact of Social Factors, Volume 18, pages 269–286.

ISBN: 0-7623-0740-4

social isolation – a consequence of uncertainty were also mentioned as major problems. In various ways, *all* my respondents talked about the uncertainty aspect inherent in chronic illness more than anything else, because this aspect of chronic illness impacts most on their quality of life, a crucial issue for all chronically ill persons.

These persons also talked about their medical problems, financial burdens and their frustrations with health care professionals, particularly about the fact that they are not getting much help from these professionals. As one of my respondents, after seeing many specialists for her condition stated: "They don't tell you right out, but you get the point . . . they can't do a thing for you." Sometimes they do "tell you right out," as another one of my respondents recalled:

> I was getting, I would say what you would call attacks which lasted about a week to ten days Then I started with a – well I had the orthopedic man, then I went to a man who just handled arthritis to double check, because this orthopedic surgeon said I had arthritis. Well, I figured why not go to somebody who all they do is just arthritis and maybe he's got something that'll help me. I went there, and he said there's nothing that can be done! It's definitely arthritis, and you have to live with it!

Almost everybody is told, one way or another, "to learn to live with it," since there is no "cure" for these chronic conditions, at our present state of knowledge. People accept that definition; what they are mostly frustrated about is the fact that they are not getting much help from health-care professionals in how to do it. There seems to be a gap between peoples' expectations about the kinds of help they feel they should be getting from these professionals and what they are getting in reality, which these chronically ill people feel is not much.

Part of the reason is that, in many cases, professionals themselves do not have all the answers, and so the ill are constantly exposed to contradictory research findings with contradictions compounded by the way medical science itself and the way health care services are delivered is constantly changing. Moreover, since most of the daily management of chronic illness actually takes place in the home and is performed by ill persons themselves, a lot of the day-to-day problems these people face are invisible to healh care providers.

Goffman (1967), states that "Information about the individual helps to define the situation, enabling others to know in advance what he will expect of them and what they may expect of him." Thus, it is crucial to take into consideration chronically ill persons' experiences and perceptions of the situation together with those of health-care professionals to get at a more realistic understanding of the total experience in living with an ongoing illness. Without that understanding, a lot of efforts and resources will be wasted without adequately meeting the needs of chronically ill people. This work discusses the uncertainty factor in living with chronic illness, the problems it creates between health-care providers and patients, and how patients try to manage the uncertainty inherent in chronic illness.

RESEARCH DESIGN

The primary data consist of focused, open-ended in-depth interviews with fifty non-institutionalized men and women between 18 and 65 years of age, suffering from a variety of chronic illnesses. The sample ranges from upper-middle class (professional) to lower class (welfare recipient) in terms of education and income. The sample is also independent of medical and institutional settings, since the purpose of the study is on everyday problems in living with chronic illness. As interviews proceeded, a special effort was made to include healthcare professionals (doctors and nurses) who also happened to be chronically ill persons. In fact, some of the most interesting observations and insights came from these people, as they experienced role reversals. Indeed, being a patient and/or being in a hospital represents only a small part of the total experience of coping with the everyday aspects of an ongoing illness. Understandably, these aspects are largely invisible to health-care professionals, and considered outside of their responsibilty, yet these concerns are of crucial importance to chronically ill persons and their families.

When I asked my respondents to describe the most problematic aspect of their illness, the concept of uncertainty emerged as the most salient characteristic of chronic illness. Uncertainty did not start out as a major focus of my research, but it turned out to be the aspect of chronic illness that seems most to affect the adaptation versus normalization processes of ill people because it undermines their quality of life.

The analysis of the data is qualitative and inductive, in the tradition of grounded theory (Glasser & Strauss, 1967).

UNCERTAINTY: A KEY PROBLEM OF CHRONIC ILLNESS

The course of events is unpredictable in chronic illness, in which remissions and exacerbations occur. In fact, chronic illness has been described by Wiener (1975) as an experience of living with chronic uncertainty. Uncertainty does not represent the total experience of chronic illness, yet it is a constant and most problematic part of it. Thus, the concept of uncertainty must be addressed in order to understand the problematic nature of chronic illnesses. Uncertainty is defined as the inability to determine the meaning of events and occurs in a situation where the decision-maker is unable to assign definite values to objects and events and/or is unable accurately to predict outcomes.

In the illness experience, uncertainty arises when: (1) people lack information about the diagnosis and seriousness of their illness; (2) they cannot make predictions as to the course of the illness and its prognosis; (3) they have ambiguity concerning the stage of their illness (e.g. how advanced is it?); (4) they lack information regarding the best and/or alternate treatments and systems of care; and (5) recently, whether they can trust their doctors not to give in to pressures from the managed care organizations to prescribe the least costly procedures rather than the more appropriate ones for their patients.

Strauss et al. (1975, 1984) used the term "trajectory" in their discussion of the uncertainty of chronic illness. They coined the term "illness trajectory" to refer not only to the physiological manifestations of a sick person's disease but also to the "total organization of work done over that course, plus the impact on those involved with that work and its organization." (1984, pp. 64). They noted that uncertain courses of illness and their trajectories tend to maximize personal and familial hardships.

Koocher (1984), reports that uncertainty regarding the duration of an illness or its oucome as being the greatest single psychological stressor for people with life-threatening chronic illness. Although, the outcome may not be the desired choice, it is preferable because it removes the uncertainty. Many researchers (Mishel, Hostetter, King, and Graham (1984); Richardson et al. (1987); Hilton (1988); Webster and Christman (1988); and Wineman (1990), all found that uncertainty led to anxiety and depression. Most of my respondents reported the experience of uncertainty as emotionally painful.

Certain illnesses and their possible developments are well known, as are the impact of their treatments. For these illnesses the course of illness is fairly predictable; their phases and the relative rate at which the phases will change can be more or less anticipated. When symptoms form a pattern, less uncertainty and less ambiguity exist about the state of the illness, so ill people have a clearer appraisal of their disease state and can monitor progress and effectiveness of their treatments (e.g. a diabetic patient periodically testing her/his blood sugar levels).

For other illnesses, such as multiple sclerosis, epilepsy, lupus, Crohn's disease, and rheumatoid arthritis, both the phases and their rates of change are very uncertain. A major characteristic of many of these illnesses is not that they are curable or incurable, but that their manifestations and trajectories are unpredictable and the management methods are uncertain. These illnesses do not follow a strictly downhill course. Flare-ups occur suddenly and, in some cases, they can also be suddenly arrested, and their trajectories can be quite calm for along time, only to burst out again. People complain that they do not know what type of symptoms will occur (i.e. pain, stiffness, swelling, and numbness) where on their body they will occur, how severe they will be, or how long they will last. Some respondents reported that they never know from one day to the next how sick they will be. But, even with relatively predictable illnesses the sequential phases do not appear and disappear in a definite time pattern. This matter of uncertainty is of utmost importance, because the efficiency of treatments and social arrangements are closely linked with the predictability of the trajectories.

Problems in family relationships have been reported as resulting from uncertainty (Mishel et al., 1984; Rowat & Knafl, 1985). Uncertainty can affect day-to-day functioning by sabotaging adjustments in the major areas of family, work, and recreation because the family's day-to-day living becomes indeterminate. For example, people suffering from rheumatoid arthritis find planning ahead impossible owing to unexpected pain and stiffness. Lazarus (1974) suggested that when an event such as chronic illness has an uncertain outcome, it is judged to be a serious threat, because the person is not able to get a clear idea of what is to occur. Shalitt (1977) stated that the ambiguity factor in uncertainty has the highest threat potential because it makes locating hazardous situations difficult and thus impedes efective coping. A respondent who thought that she could control her illness by avoiding potentially harmful activities and sitautions and still find herself in trouble, explained:

> I could just be sitting at home doing absolutely nothing and all of a sudden, I'll start having trouble breathing. It happened to me one time, when I was sitting at home, watching television and all of a sudent I started feeling like I couldn't breathe. It could have been for any number of reasons, but . . .?

Mishel (1981) also noted that uncertainty hampers the formation of a plan of action, limiting individuals's abilities adequately to evaluate their situation, and according to Mishel and Sorenson (1991) it also weakens their sense of mastery over a situation. Murdaugh and Mishel (1991), found that uncertainty weakens internal locus of control; Braden (1990) reported that uncertainty reduced their sense of learned resourcefulness among chronically ill persons suffering from rheumatoid arthritis, which in turn lessened their ability to help themselves.

Indeed, we know from the social psychological literature that individuals define situations through some cognitive process and that, based on this evaluation, they determine the appropriate management strategies (Lazarus, 1966; Mechanic, 1974; Chirisman, 1977; Lazarus & Cohen, 1977; Shalitt, 1977). A major and erroneous assumption made in most of these discussions of cognitive evaluation is that the person in question has complete information about the situation being appraised. Yet, in most cases of chronic illness there is consideratble uncertainty and lack of information. For instance, the symptoms of a disease, although routinely seen by physicians and known to them to be non-life-threatening, may nevertheless be unfamiliar, unexpected, confusing and frightening to patients. Physicians who tell patients to relax, not to worry, to reduce their activities, or to call back later if the symptoms persist, commonly known as the "take two aspirins and call me in the morning diagnosis", without giving any explanations for these symptoms or any rationale for their advice, only add to the uncertainty and fear that most of these patients feel. A respondent, recalling a very frightening experience, said:

> I was having an attack at the time. The first thing he did was give me a shot in the arm. You know, I just asked him, I go: What exactly is going on? He goes: You have Asthma! He didn't really say much. He didn't tell me that much. He just gave me a shot and told me to go get a prescription filled and that was about it. And, I'm not even sure if he gave me a pamphlet on it. . . . He just told me that I have it and that's about it.

There is a certain amount of uncertainty and stress inherent in living with a chronic illness, but when people have to make their decisions on the basis of limited or incorrect information, it creates additional and unnecessary uncertainty and stress that makes a difficult situation even worse. Chronically ill people, like most of us, are not experts in the healthcare field, and are usually given limited or no information about their illness, its treatment methods and alternatives, by those who treat them, under the assumption that these ill people would not be able to comprehend all these complex issues. Yet, these chronically ill people have to make crucial decisions affecting all aspects of their lives based on insufficient information, at a time when they are most vulnerable and anxious. Therefore, providing chronically ill persons with sufficient information about their illness and available treatment alternatives is one important means of reducing uncertainty, stress, and need to search for causal explanations they all experience. Lack of confidence and trust in the physician results in high levels of uncertainty. Mishel (1988), found that confidence and trust in their physician explained 35% of the variance in uncertainty because the physician is a source of information about causes and consequences of symptoms. The more credible the physician, the less the uncertainty.

When events are recognized as familiar, less uncertainty exists about the complexity of treatment and the system of care. Knowledge of the usual symptom display allows the sufferer a degree of control because it allows comparisons for monitoring changes during treatment. Direct inquiry by healthcare providers to elicit sufferers' concerns about their illness and response to these concerns improves their understanding of the symptom pattern. Thus, effective communication between health-care providers and the chronically ill on the causes and consequences of symptoms serves to alleviate some uncertainty and fear and can be an important factor in determining the way ill persons evaluate their situations. As one respondent stated:

> Some people might not be happy with a doctor that will tell you every thing; my doctor will, and he's been very supportive; he's honest. He tells you there is no miracle cure for it, . . . he says if you have any questions about it just ask me, and I'll find out any answers that are available. Matter of fact, the last time I saw him he gave me a book he wanted me to look over I explained to him, I said, well . . . he realized I can't read too well, so I says what will happen is my wife will read it to me . . . I picked out a chapter that I was interested in at the time and my wife read the whole book. . . . He said let him know what you guys think of it, and we told him that we thought it was a fairly good description of it and it didn't hold anything back.

In fact, persons who recently have been diagnosed as having a long-term chronic illness are profoundly fearful and uncertain about many things. An environment that was formerly at least tolerable has now become unpredictable and threatening, bringing the former assumptions and beliefs these persons have about the self and the environment into question. These people are thus confronted with a web of fears, including fear of pain, of recurrence, of progressive deterioration, of dying, and most likely fear of dependency on medications and other people. A respondent expressed these feelings by:

> I don't want to be dependent on it, you know, I can't see myself doing that for the rest of my life, although, I know someone with asthma, it's very possible. My uncle had asthma and he was dependent on medication for all of his life, but I don't like to be dependent on things, or on people either. Not too much.

Most people also worry about whether the doctors have been honest with them, whether they are receiving the best care available, and also worry about how their illness will affect their families, friendships and work situations. The same respondent above was fighting against dependency on medications but was forced to give in because she was worried about her relationships and even more reluctant to be dependent on other people

> I kept having to stop and you know, trying to catch my breath and go a few more feet and then stop because I didn't have the medicine to help me out. . . . I didn't want to have to put them through it. I'm like, I'm really sorry to be doing this to you guys and everything, 'cause they'd take me home and make sure that I was okay, so now I just always carry it [the medicine] with me. So, I guess I'm not so much a burden to everyone else.

These people are forced to contend with a variety of physical changes and problems, which may include pain, disfigurement, nausea, energy loss, malodors. Clearly such changes are deeply unsettling and can have a profound effect on persons' self-concepts and consequently on their relationships.

What is ironic is that within the context of this uncertainty about the changing environment and self, chronically ill individuals are often called upon to function more competently and more rationally than ever before in making the many decisions that confront them. These decisions range from questions about which physician to contact, what hospital to enter, and what treatment to have, to questions about what to tell spouses, children, friends and co-workers. Thus, most ill people are at first overwhelmed by the number and complexity of decisions to be made. While these decisions are similar to those that people make in everyday life, they are linked to graver consequences, and chronically ill persons are emotionally and physically less able to tackle them. In addition, a large number of these decisions concern matters with which most people have no prior expertise or experience. The above mentioned respondent, finally realizing how intrusive her illness really was, said:

> It's really strange, cause like you know, when they said I had asthma, I never thought of that as really a chronic illness, but you know, I guess it is! I don't usually think about it that much. I just cope with it from day to day. I just make the best of it I can.

As a matter of fact, there is considerable ambiguity within the field of medicine itself about how to treat many chronic conditions since most of them, by definition, cannot be cured at this stage of our medical knowledge. Unlike the case of acute illness, medicine has made limited progress in the cure and treatment of chronic illness, which is not susceptible to a single-cause approach. As far as we know, at this state of our knowledge, many chronic illnesses seem to have no known clear-cut physical cause. Consequently, the rise and prevalance of chronic illness has led to the development of multi-causal theories of disease, including consideration given to differences in life-styles and socio-environmental stress in causation of disease. Unfortunately, there may be no clear-cut answers either to some of the problems facing afflicted individuals.

The literature pertinent to how physicians deal with uncertainty has received increasing attention in recent years (Bosk, 1979; Fox, 1980; Light, 1980). First explored by Renee C. Fox (1957), this dimension of medical training creates a double bind for physicians because uncertainty pervades their work and they must be open to it; yet their license and mandate from society presumes they know how to take care of problems which ordinary persons are uncertain how to address. It is interesting to note that Oliver Wendell Holmes wrote to a young physician, "Let me recommend to you, as far as possible, to keep your doubts to yourself, and give the patient the benefit of your decision" (In Mumford, 1970, p. 163). Indeed, there is a lot of evidence in the health and illness literature focusing on doctor/patient relationships that this advice is well followed. For instance, until recently, patients were not even allowed to see their own charts, and most of the time, they still do not have access to vital information about their conditions. When, patients ask to see their charts they are met with suspicion and hostility, because the assumption is made that they are gathering material to bring legal action against their doctors or the hospital.

Whatever the reason, most of the time, ill persons are not told about treatment alternatives. For instance, some surgeons are still performing mastectomies when in fact many studies in the last twenty years have demonstrated that less disfiguring procedures such as removing only the malignant lesion and radiotherapy produces the same survival rates as removing the whole breast, which can have devastating psychological effects on some women. On the other hand, some women are not told of the side effects and consequences of radiotherapy, so they cannot decide rationally which operation would be best for their particular circumstances. As another example: there are almost as many different kinds of operations for treating heart conditions as there are prominent heart surgeons in our society. Some of these procedures are less debilitating, yet seem to show the same survival rates as the more drastic measures. But since chronically ill persons are not usually informed about all these different kinds of operations available, it is often impossible for them to feel confident that the chosen treatments are the best ones. One of my respondents who had been a general practitioner for over twenty years before he "suddenly collapsed into a coma because of kidney failure" and found himself a reluctant "patient," exclaimed:

> I never realized what patients have to go through! It's been quite an education. They treat you like morons, they come and go, do things to you and they don't tell you anything. Yes, even me!

In fact, people customarily get the treatments their particular physician or surgeon is trained in or familiar with, without being fully informed about alternative procedures regardless of their effectiveness or cost.

Of course, there are many reasons why physicians might communicate in a manner that leaves patients confused and frustrated: (1) they might not know exactly what to do, or be able to predict outcome; (2) the known treatments may have been exhausted with no positive results; or (3) as a result of treating some symptoms, additional and unexpected problems ("iatrogenic disease") arise.

Fox (1989) notes that "the mode of thought on which modern medical science and research are based is an emergent, eclectic product of the numerous fields of biological, physical, and behavioral science on which it draws. Biology is its cornerstone, but the phenomena with which medicine deals are both too complex and too individuated to be confined to a purely biological analysis, or to any single explanatory theory." (p. 183) She further notes, that "By its very nature, . . . scientific research is fraught with uncertainty, with seemingly aleatory events, and with limitations in knowledge – including those that are inherent to scientific ways of knowing. . . . medical researchers experience these problems in a particularly acute and often painful way. They are faced with the immediate consequences for health and illness, life and death, of what they do not know or understand, and cannot control." Indeed, doctors complain that they never expected the degree of uncertainty inherent in medicine when they chose it as a career and that they have problems dealing with the stress the uncertainty exerts.

In her earlier research, Fox (1959) has documented the agony of well-intentioned and sincere physicians who vigorously treat disease only to find that the treatment itself creates unresolvable complications, seen as medical failures by patients and physicians alike. Indeed, failure is difficult to acknowledge by anyone, and telling patients and their families that nothing more can be done cannot be easy.

STRATEGIES USED TO DEAL WITH UNCERTAINTY

Since the uncertainty factor in chronic illnesses makes life more unpredictable for many chronically ill persons and for their families, these persons avoid many potentially embarrassing situations. For example, one respondent recalled that she had disapointed her family and friends a couple of times by offering to make Thanksgiving dinner or having Christmas at her house, because:

> I was feeling pretty good at the time and I wanted to do it . . . they were always helping me. I love having people around, they love my cooking. But when I started doing it . . . it was too much, I got so sick. One time they had to go out and eat out. . . . I think, it was Thanksgiving . . . Another time they had to do all the work, and I could not even help, or eat with them, and I couldn't even sit at the table. They said it was O. K. but I felt like. . . . So, I just don't plan on it, don't offer anymore. I am afraid it's gonna happen again, so I just avoid it.

In fact, since most people, chronically ill or not, would not knowingly put themselves in potentially embarrassing situations, chronically ill persons, being in greater jeopardy for such occurrances, also use the strategy of limiting their contacts to persons in similar circustances.

How do patients cognitively deal with uncertainty? I thought that one of the most fascinating questions about the chronically ill had to be how they evaluate their own adjustment to their condition. Indeed, because of the uncertainties they face and because their sense of self is threatened, many chronically ill persons probably experience intense needs both to clarify what is happening to them and to be supported and reassured by others. Thus, I hypothesized that if people think they are dealing badly with their illness it will be an added strain on what may be an already difficult existence, whereas if these individuals feel they are adjusting well, that belief can be a source of satisfaction and pride.

As Schachter (1959) stated, people often seek out other people for purposes of social comparison or reassurance when doubts arise about their own-experiences. As far as chronic illness is concerned, this process of social comparison, although essential, can have serious consequences at times, because some people might exacerbate a problem by persisting in behaviors condoned or encouraged by others; they might postpone treatment until it is too late, having been told by others that treatment is either hopeless or unnecessary; or they might discontinue an effective therapy, impressed by the negative reactions of their families and friends.

Nevertheless, whatever the consequences, chronically ill people, like all individuals, do indeed compare themselves to others, yet they do it somewhat differently. The literature on social comparison demonstrates that in general individuals compare themselves with people who are doing slightly better than they are. According to Festinger (1954), although one comes off as psychologically disadvantaged, one may gain useful information from the upward comparison.

There are indications however, that some people, including chronically ill people, appear instead to be making downward comparisons, so they come off assome what psychologically advantaged and informed. Some research on social comparison indicates that this pattern of downward comparisons is particularly likely to occur under conditions of threat (Hakmiller, 1966; Wheeler et al., 1969).

This finding might suggest that these individuals may be selecting their comparisons to preserve their self-esteem, rather than letting self-esteem be determined by those readily available for comparison. Such downward comparisons as "I'm really much better off than she/he is . . ." can bolster these peoples' own sense of self, and might convince them that they are adapting well to their difficult circumstances.

My data strongly support this finding; of all the people I interviewed, not a single person seemed to know anybody suffering from the same illness who was better off than themselves. All these people either said they knew of no one in the same circumstances or they told me horror stories about similar others. For example, when asked if she knew other people with the same illness, a respondent

> . . . my brother-in-law got married, and his wife's father had M. S., I never met the man because he was bedridden in the hospital in Park Forest. In Park Forest? I never could keep them straight, out south somewhere, but they never said too much about him. I just was

> always like: well, why couldn't he come to the wedding? At least, they could have done that for him. But, at the time I didn't realize he was so bad. They just couldn't put him in a wheelchair and bring him to the wedding. Ya know, when you first get these things, and you don't wind up flat on your back, you don't realize what the worst is.

Thus, it seems for reasons of reassurance and/or preservation of self-esteem, chronically ill people often look to the people in similar circumstances to gather information about and evaluate their conditions, their feelings, their behaviors in order to reduce their illness-related uncertainties, and particularly, to evaluate their own performances. But, to bolster their self-esteem and make themselves feel and look good, they seem to select or remember only "worst scenario" examples and thus to compare themselves only to people who are worse off than themselves. A respondent suffering from kidney failure remarked:

> Well, some of the people are much worse off than I, not all of them, but there is one woman who has diabetes and her legs are all swollen and, I don't like puss. They're all bandaged up and like puss or something -blood and everything comes through the bandages and I mean she is in terrible shape. And, there's another one, she had diabetes and also kidney failure . . . and they had to cut her toes off and then they had to take her eye out. I mean when you look at that, you think I'm not that bad off!

Since of all the people I interviewed, as noted, not a single person admitted knowing another individual suffering from the same illness who was doing better or coping better than themselves, it made me think that these people either purposely ignored those who are doing better under similar circumstances, or more likely, they chose to remember only those people who are doing much worse. For example, a respondent told me that having arthritis was very common among people in his line of business – plumbing, construction work. However, when pressed to come up with a specific example, he said: "I've known several people in the business, at this point, I don't know." He later remembered:

> My nephew's wife's mother, his mother-in-law I would say, had arthritis in her hands and they operated on all of of them, and took out the bone or the calcium and now she can move her fingers like a person. . . . Well the ones that I have [known] I guess have rheumatoid arthritis, that's the worst, when the bones are affected!

Such downward comparisons as "I'm really much better off than she is . . ." (a comment made by a respondent comparing herself to a friend in the same situation), do seem to bolster her sense of self. Indeed, this respondent who is severely suffering from multiple sclerosis, yet leading an active social life as wife, mother and travelling companion for her husband, proudly declared:

> I have a friend who – we met through a mutual person – same age as I am, has four children, and in five years she was in a wheelchair, could not use her legs at all, and could

barely use her arms. She was a nurse, she had four young children, and the same situation, identical. She was an ambitious person . . .?

Even if they don't know a specific individual for comparison, chronically ill persons still seem to think that they are doing better, coping better or in general are better off than others in the same circumstances. For instance, another respondent also suffering from arthritis, describing at length her various difficulties, still concluded by saying:

> But, it's such a severe pain, I just, you know, I had two cesareans, I had a hysterectomy, I never ever had pains like this and it's so hard to deal with! If I type, I can't type no longer than an hour at a time. I can't clean house like I used to. I can't bowl. . . . I went to the health club three times a week. The first year when I got off drugs, I gave it up completely 'cause half the times I couldn't get my arms over my head to take my exercises. . . . when I got this inmy hand, I had to take all the fluid out of my hand, and fill it with Cortizone because I couldn't straighten it up. And, I notice now that sometimes when I like sit for too long and I go to get up, it's always my fear, all the way up to ten minutes until I can get. I'm a morning person. I used to jump out of bed, you know, loving mornings. Now, sometimes when I go to jump out of bed, I can't move out of bed, so I sorta just roll out of bed. But, mornings for me aren't as bad as they are for most arthritics! Me, I don't get as stiff!

Furthermore, I noticed that chronically ill people whose adaptations to the illness, or normalization efforts, appear to be more successful than those not as able to adapt well (by their own definitions), choose dimensions of self-evaluation on which they can feel good. No matter how bad things are, these people seem to find an aspect of their situation or a personal characteristic to make themselves feel better. As the respondent suffering from multiple sclerosis, still comparing herself to her friend, further elaborated:

> Yes, she got it [the illness] at the same age as me. Now, why did she so quickly wind up in a wheelchair? And, really, I didn't wind up with anything. I was still able to keep going. Of course my mother says I'm so stubborn, I would not stay in a wheelchair. Because I was in a wheelchair for three months that I was in the hospital. . . And, when I came home I was in a wheelchair for a couple of months, but I wasn't about to stay there, if I was able to manage at all. So, as soon as I was able to grasp the cane so I'd have support, I was up! Or, maybe I had more determination than some people. I don't know.

Of course, some of the beliefs reflected in these processes may be illusionary, based less on fact than hope; yet for people trying to cope and learn to live with ongoing and incurable illness, hope and illusions pave the way for constructive action that will enable them to go on with their lives. Sometimes, the hopeful illusion continues even when it is not really justified. At the time of the interview, the above quoted respondent was in a wheelchair herself.

In short, the problem of uncertainty for people with chronic illnesses is recognized, but few studies have addressed this most important issue. As

Parsons (1951) noted, distress or at least uneasiness with the uncertainty inherent in chronic illness is to be expected in a society that emphasizes what he called "mastery" of health.

IMPLICATIONS FOR HEALTHCARE

Assuming we were medically doing all we could and trying as hard as we might, individual efforts would not always be successful in solving the problems associated with chronic illness. These medical efforts, as crucial as they are, need to be supplemented with effective structural changes in how health care is delivered and with social interventions targeted to particular problems experienced by chronically ill individuals. A big part of the problem is the lack of fit between the needs of the chronically ill and the trends and practices in acute care facilities. Clearly, the medical model, with its individualistic and reductionistic thinking, and the home care system that has fashioned itself on this model, have limitations for chronic illness care. In fact, it is inconsistent with the needs of a society with many chronically ill members, some of them aged and frail, who need supportive services to maintain their lifestyles and their independence. Chronically ill individuals need a health care delivery system sensitive to patient experiences and perceptions. As it is, the needs of most chronically ill people are not being met by the present health care structure. We need a system in which patients are taught how to care for themselves, caregivers are taught how to provide care, and both learn how to access needed services.

In a society where most people, men and women, have to work for wages outside their homes, there is virtually no one available to take care of sick and elderly people on a daily basis. Most chronically ill persons, particularly younger people, have no choice but to work and at the same time manage their illnesses in the context of their lives. The difference between chronically ill people and so-called healthy people is that the former have additional burdens; their goals and desires to remain active and involved in their valued life activities do not change after they become ill. In fact, I found that the actual severity of the illness appears to be unrelated to a person's ability to remain involved in valued pursuits. As my respondents indicated, people will go to great lengths to maintain their desired lifestyles in spite of their illness. Thus health care professionals must take the social and psychological factors related to living with long-term illness into consideration when treating such patients. A health care system that focuses only on medical aspects, ignoring peoples' concerns with issues about their quality of life, is not much help to these people.

Traditionally, there was also a notable blurring of definitions among conditions of chronic illness, disability, and long-term care. Presently we recognize that not all individuals who are chronically ill are functionally disabled, many disabled persons do not require any significant amount of health care, and most chronic illnesses do not require long-term care in an institution, even though these conditions cannot be cured. Furthermore, health care delivery systems are constantly changing, and since most of the daily management of chronic illness actually takes place in the home and is performed by ill persons themselves, a lot of the day-to-day problems these people and their families face are invisible to health care providers. Therefore, it is crucial to take into consideration chronically ill peoples' experiences and perceptions of the situation together with those of health care professionals, in order to get at a more realistic understanding of the total experience of living with an ongoing illness.

In fact, identification of a medical problem requires at least three steps: (1) the prospective patient must recognize that there is a problem and seek medical help, (2) the medical personnel must make a correct diagnosis, and (3) medical personnel must convey this information to the patient and make sure the patient understands what is involved, so that, ideally, caregiver and patient together can decide on and implement a treatment plan.

The traditional "medical model" suggests that there is a recommended treatment or procedure that will cure or alleviate almost any illness. Yet the success of a treatment often depends on the willingness of the patient, especially the chronically ill patient, who most of the time is not under the direct supervision of health care personnel, to comply with the prescribed regimen – to take the pills, show up for chemotherapy or dialysis treatments, observe dietary restrictions, and the like. An important task of the medical practitioner, then, is to convey information to the chronically ill person – information that presumably will affect the person's compliance. Krantz, Grunberg and Baum (1985) found that in general, the more satisfied patients are with the quality of their communication and interaction with their physicians, the more likely they are to comply with the recommended treatments. Yet too often physicians volunteer too little or no information and the patient asks too few questions, with the result that the potential treatment program encounters real problems (DiMatteo & DiNicola, 1982), with negative consequences for care and management of illness. For example, a doctor who maintains contact with families rather than just individuals can more easily understand the many different ways in which people think, feel, and act and the importance of values in medical decision making. Indeed, a doctor who treats chronically ill people, regardless of specialty, needs to be a bit of a family doctor as well.

My data shows that patients are angry or disturbed about what they perceive as "noncaring and judgmental attitudes" of their caregivers, particularly their doctors, without realizing the structural and financial impediments under which individual doctors have to practice medicine. Part of the problem is that more and more chronically ill patients are seen as creating a financial and caseload crisis in hospitals and in an overburdened health care system in general. This group of patients are unpopular with both physicians and hospital administrators. They also strain the already diminishing supply of nurses, because their health care and related needs are constant. Some of these needs are beyond the training, capabilities, and responsibilities of medical professionals.

To be fair, as chronic illness has become our major health problem, its physical, social, and psychological consequences have been increasingly recognized, and a number of specific programs have been initiated to deal with particular problems. Yet at the present time, these programs are not systematically coordinated or widely available to the majority of chronically ill persons. As Relman (1990) states:

> What our health care system needs now is not more money, but different incentives and a better organization that will enable us to use available resources in a more equitable and efficient manner to provide necessary services for all who need them. We can afford all the care that is medically appropriate according to the best professional standards. We cannot afford all the care a market-driven system is capable of giving (p. 913).

A chief goal for health care professionals, social psychologists, and health psychologists in the coming years, then, should be to help develop programs and cost-effective interventions to improve the quality of life of chronically ill persons and their families.

A first task should be the coordination, identification, and mobilization of existing services within a community so that chronically ill persons and their families can be referred to them when appropriate. A second task should concentrate on making sure that the needed services are available in every community. A third and more important task should involve instituting a standardized and structured system of health care by which an initial assessment of needs in all domains of a patient's life is made right after the patient is diagnosed with a particular chronic illness, during the initial acute period of that illness.

Of course, the initial assessment must be supplemented with regular needs assessment over the long term to identify potential problems, such as depression, anxiety, gaps of knowledge, and difficulties with compliance, before they fully disrupt a patient's life. Systematic and regular assessment, in turn, can help define further the needs of the chronically ill, pointing to new services that may need to be developed. The main purpose of such a system would be to help develop cost-effective interventions to prevent costly and disruptive acute episodes and to improve quality of life for chronically ill people and their families. A network of home and community-based care as an alternative to hospital care can bring about healthier patients, fewer problems, fewer hospitalizations, and fewer drugs – with a more cost-effective result (Royer, 1998).

REFERENCES

Bosk, C. L. (1979). *Forgive and Remember: Managing Medical Failure*. Chicago: University of Chicago Press.

Braden, C. J. (1990). A Test of the Self-Help Model: Learned Response to Chronic Illness Experience. *Nursing Research, 39*, 42–47.

Chirisman, N. (1977). The Health Seeking Process: An Approach to the Natural History of Illness. *Culture, Medicine, and Psychiatry, 1*, 351–377.

Festinger, L. (1954). A Theory of Social Comparison Process. *Human Relations, 7*, 117–140.

Fox, R. C. (1957). Training for Uncertainty. In: R. K. Merton, G. C. Reader & P. Kendall (Eds), *The Student-Physician: Introductory Studies in the Sociology of Medical Education* (pp. 207–41). Cambridge, MA: Harvard University Press.

Fox, R. C. (1959). *Experiment Perilous*. New York: Free Press.

Fox, R. C. (1980). The Evolution of Medical Uncertainty. *Milbank Memorial Fund Quarterly/ Health and Society, 58*(19) (Winter).

Fox, R. C. (1989). *The Sociology of Medicine: A Participant Observer's View*. Englewood Cliffs, NJ: Prentice Hall.

Glaser, B. G., & Strauss, A. L. (1967). *The Discovery of Grounded Theory*. Chicago: Aldine.

Goffman, E. (1967). *Interaction Ritual: Essays on Face-to-Face Behavior*. New York: Doubleday.

Hakmiller, K. L. (1966). Threat as a Determinant of Downward Comparison. *Journal of Experimental Social Psychology Supplement, 1*, 32–39.

Hilton, B. A. (1988). The Phenomenon of Uncertainty in Women with Breast Cancer. *Issues in Mental Health Nursing, 9*, 217–238.

Koocher, G. P. (1984). Terminal Care and Survivorship in Pediatric Chronic Illness. *Clinical Psychology Review, 4*, 571–583.

Lazarus, R. (1966). *Psychological Stress and Coping Process*. New York: McGraw-Hill.

Lazarus, R. (1974). Psychological Stress and Coping in Adaptation and Illness. *International Journal of Psychiatry in Medicine, 5*, 321–333.

Lazarus, R., & Cohen, J. (1977). Environmental Stress. In: I. Altman & J. Wohlwill (Eds), *Human Behavior and Environment* (vol. 2). New York: Plenum.

Light, D. (1980). *Becoming Psychiatrists: The Professional Transformation of Self*. New York: Norton.

Mechanic, D. (1974). Social Structure and Personal Adaptation: Some Neglected Dimensions. In: G. Coelho., D. Hamburg & J. Adams (Eds), *Coping and Adaptation*. New York: Basic Books.

Mishel, M. H. (1981). The Measurement of Uncertainty in Illness. *Nursing Research, 30*, 258–263.

Mishel, M. H. (1988). Uncertainty in Illness. *Image: Journal of Nursing Scholarship, 20*, 225–232.

Mishel, M. H., Hostetter, T., King, B., & Graham, V. (1984). Predictors of Psychosocial Adjustment in Patients Newly Diagnosed with Gynecological Cancer. *Cancer Nursing, 7*, 291–299.

Mishel, M. H., & Sorenson, D. S. (1991). Uncertainty in Gynecological Cancer: A Test of the Mediating Functions of Mastery and Coping. *Nursing Research, 40*, 167–171.

Mumford, E. (1970). *Interns: From Students to Physicians*. Cambridge, MA: Harvard University Press.

Murdaugh, C. L., & Mishel, M. H. (1991). *Predictors of Quality of Life in Patients Who Undergo Heart Transplantation*. Paper presented at the American Heart Association Meeting. Anaheim, CA.

Parsons, T. (1951). *The Social System*. Glencoe, IL: Free Press.

Richardson, J. L., Marks, G., Johnson, C. A., Graham, J. W., Chan, K. K., Selser, J. N., Kishbaugh, C., Barranday, Y., & Levine, A. M. (1987). Path Model of Compliance with Cancer Therapy. *Health Psychology, 6*, 183–207.

Rowat, K. M., & Knafl, K. A. (1985). Living with Chronic Pain: The Spouse's Perspective. *Pain, 23*, 259–271.

Royer, A. (1998). *Life with Chronic Illness: Social and Psychological Dimensons*. Westport, CT: Praeger Publishers.

Schachter, S. (1959). *The Psychology of Affiliation*. Stanford, CA: Stanford University Press.

Shalit, B. (1977). Structural Ambiguity and Limits to Coping. *Journal of Human Stress, 3*, 32–45.

Strauss, A. L., & Glaser, B. (1975). *Chronic Illness and the Quality of Life*. Saint Louis: C. V. Mosby.

Strauss, A. L., Corbin, J., Fagerhaugh, S., Glaser, B. G., Maines, D., Suczek, B., & Wiener, C. L. (1984). *Chronic Illness and the Quality of Life* (2nd ed.). Saint Louis: C. V. Mosby.

Webster, K. K., & Christman, N. J. (1988). Perceived Uncertainty and Coping Post Myocardial Infraction. *Western Journal of Nursing Research, 10*, 384–400.

Wheeler, L., Shaver, K. G., Jones, R. A., Goethals, G. R., Cooper, J., Robinson, J. E., Gruder, C. L., & Outzine, K. W. (1969). Factors Determining the Choice of a Comparison Other. *Journal of Experimental Social Psychology, 5*, 219–232.

Wiener, C. L. (1975). The Burden of Rheumatoid Arthritis. In: A. L. Strauss & B. Glaser (Eds), *Chronic Illness and the Quality of Life* (pp. 71–80). St. Louis: C. V. Mosby.

Wineman, N. M. (1990). Adaptation to Multiple Sclerosis: The Role of Social Support, Functional Disability and Perceived Uncertainty. *Nursing Research, 39*, 294–299.

www.ingramcontent.com/pod-product-compliance
Ingram Content Group UK Ltd.
Pitfield, Milton Keynes, MK11 3LW, UK
UKHW020845290926
15470UKWH00004B/1557

9 780762 307401